About Island Press

Since 1984, the nonprofit Island Press has been stimulating, shaping, and communicating the ideas that are essential for solving environmental problems worldwide. With more than 800 titles in print and some 40 new releases each year, we are the nation's leading publisher on environmental issues. We identify innovative thinkers and emerging trends in the environmental field. We work with world-renowned experts and authors to develop cross-disciplinary solutions to environmental challenges.

Island Press designs and implements coordinated book publication campaigns in order to communicate our critical messages in print, in person, and online using the latest technologies, programs, and the media. Our goal: to reach targeted audiences—scientists, policymakers, environmental advocates, the media, and concerned citizens—who can and will take action to protect the plants and animals that enrich our world, the ecosystems we need to survive, the water we drink, and the air we breathe.

Island Press gratefully acknowledges the support of its work by the Agua Fund, Inc., Annenberg Foundation, The Christensen Fund, The Nathan Cummings Foundation, The Geraldine R. Dodge Foundation, Doris Duke Charitable Foundation, The Educational Foundation of America, Betsy and Jesse Fink Foundation, The William and Flora Hewlett Foundation, The Kendeda Fund, The Forrest and Frances Lattner Foundation, The Andrew W. Mellon Foundation, The Curtis and Edith Munson Foundation, Oak Foundation, The Overbrook Foundation, the David and Lucile Packard Foundation, The Summit Fund of Washington, Trust for Architectural Easements, Wallace Global Fund, The Winslow Foundation, and other generous donors.

The opinions expressed in this book are those of the author(s) and do not necessarily reflect the views of our donors.

Salvage Logging
and Its Ecological Consequences

Salvage Logging

AND ITS

Ecological Consequences

David B. Lindenmayer
Philip J. Burton
Jerry F. Franklin

ISLANDPRESS

Washington · Covelo · London

Lindenmayer, David.
  Salvage logging and its ecological consequences / David B. Lindenmayer, Philip J. Burton. Jerry F. Franklin.
     p. cm.
  Includes bibliographical references and index.
  ISBN-13: 978-1-59726-402-0 (cloth : alk. paper)
  ISBN-10: 1-59726-402-4 (cloth : alk. paper)
  ISBN-13: 978-1-59726-403-7 (pbk. : alk. paper)
  ISBN-10: 1-59726-403-2 (pbk. : alk. paper)
  1. Logging—Environmental aspects. 2. Salvage (Waste, etc.) 3. Forest ecology. I. Burton, Philip J. II. Franklin, Jerry F. III. Title.
  SD538.L73 2006
  577.3'24—dc22   2007050708

Printed on recycled, acid-free paper

Manufactured in the United States of America

10 9 8 7 6 5 4 3 2 1

Keywords: salvage logging, salvage harvesting, wildfire, fire ecology, natural disturbance regimes in forests

DEDICATION

To the memory of Jordan (Jordy) S. Tanz (1953–2006), a man devoted to his family, to experiencing the wilderness by canoe, to lifelong education, to forests and sustainable forestry (photo courtesy of Jeff Tanz).[1]

---

[1]See http://www.cortex.ca/teatan.html

CONTENTS

Large-scale natural disturbances such as wildfires, insect attacks, hurricanes, cyclones, and floods are commonplace around the world. Climate change is predicted to make some of them more frequent and others more intense or more widespread. The management of ecosystems, particularly forests, after natural disturbance often includes salvage logging to recover economic value or prepare sites for some form of revegetation.

Salvage logging is a highly controversial topic. Are the ecological impacts of cutting "green" timber from undisturbed forests greater than the salvage logging of naturally disturbed areas? Or does salvage logging confer a "double hit" to already stressed ecosystems? Are naturally disturbed young forests rare ecosystems, and should they be assigned special protection similar to that often afforded old-growth forest? Should naturally disturbed areas be subject to tree planting to speed reforestation, or should they be allowed to recover naturally?

Issues associated with salvage logging are not well understood either by the general public or by many foresters, ecologists, and other scientists and land managers. Moreover, not enough science on salvage logging has found its way into policymaking to better guide whether salvage logging should or should not take place, and where it does occur, how such operations might be modified to limit potential negative effects.

In this book, we have tried to provide new insights to better inform policymaking and on-the-ground prescriptions for salvage logging. These are ambitious aims, but we hope that at least we have made a reasonable contribution to the debates on salvage logging. We also hope that there might be greater appreciation of the many kinds of ecological values associated with forests and other ecosystems subject to major natural disturbances. We invite feedback on the book, particularly recommendations that would improve future salvage logging operations.

David Lindenmayer
Philip Burton
Jerry Franklin
October 2007

ACKNOWLEDGMENTS

This book originated at a campfire discussion at Round Lake near Smithers in British Columbia, Canada. Two of us (DBL and PJB) were discussing issues associated with salvage logging of forests after large-scale disturbances, such as insect attack and fire. That discussion followed from the Doug Little Lecture Series sponsored by the University of Northern British Columbia and Canadian Forest Products Ltd., which had allowed DBL to travel to interior British Columbia to view firsthand the extensive salvage logging of subboreal forest in the wake of an outbreak of mountain pine beetle.

DBL would like to thank a wide range of colleagues for collaborative work on forests and salvage logging. These include Reed Noss, Keeley Ough, Dave Perry, Mac Hunter, Fiona Schmiegelow, David Foster, Mike McCarthy, Ross Cunningham, Charles Tambiah, and Dominick Della-Sala. Past work on salvage logging, including a special section in the journal *Conservation Biology* (on which part of chapter 3 is based), was supported by the Wilburforce Foundation, championed in particular by Gary Tabor (formerly of that foundation). Dr. Rebecca Montague-Drake, Rachel Muntz, and Nicki Munro helped DBL gather much of the literature used to write this book.

PJB would like to thank Nancy-Anne Rose, Darin Brooks, Doug Beckett, Albert Nussbaum, Werner Kurz, and Mart-Jan Schelhaas for help compiling statistics and other information. Support from DBL and the Pacific Forestry Centre of the Canadian Forest Service (Natural Resources Canada) for travel to Australia and to Seattle is gratefully acknowledged. The willingness of the Ecosystem Science & Management Program at the University of Northern British Columbia to sponsor two international seminar speakers on salvage logging impacts in 2005 provided a crucial foundation for this book. Colleagues in forest research across northern British Columbia are thanked for generously sharing their data, perspectives, and insight.

JFF acknowledges the valuable continuing dialogue with K. Norman Johnson, which has been critical to informing his perspectives on salvage logging. The work at Mount St. Helens with numerous colleagues, most particularly Fred Swanson, Jim Sedell, Charlie Crisafulli, and Peter Frenzen, has provided a critical foundation. JFF acknowledges recent discussions with members of his laboratory, including Mark Swanson, Andrew Larson, James Freund, and Linda Winter, as well as work on forest succession with Charles Halpern.

Comments by Rachel Muntz and a range of colleagues greatly improved earlier versions of our manuscript.

Chapter 1

The Challenge of Large-scale Natural Disturbances and Salvage Logging

Wildfires and other large-scale forest disturbances figure prominently in forested regions of the world. These disturbances (nominally "forces of nature") are events with a high public profile because of their threats to human life and values, occurrence in famous places, and their spatial scale. Peri-urban areas of major metropolitan regions, such as Los Angeles and Sydney, regularly experience intense wildfire. There have been many widely publicized and large-scale disturbances, such as the eruption of Mount St. Helens in 1980, the Yellowstone fires in 1988, well-known hurricanes such as Hugo (1989) and Katrina (2005), the Columbus Day windstorm of 1964, and the Boxing Day tsunami of 2004.

Large natural disturbances have occurred recently in many regions of the world. Wildfires burned nearly 10 million hectares in a single year (1997–98) in Indonesia (Food and Agriculture Organization 2001), 7 million hectares in 1995 in Canada (Canadian Council of Forest Ministers 2007), 3 million hectares in the 2002–3 fire season in southeastern Australia (Australian Bureau of Statistics 2004), 2.7 million hectares in 1997 in Mongolia (Food and Agriculture Organization 2001), 2 million hectares in 2000 in the United States, and 800,000 hectares in Mexico in 1998 (Food and Agriculture Organization 2001). European forests experience damage to 35 million cubic meters of wood annually (table 1.1), and impacts of the agents of disturbance are increasing (Schelhaas et al. 2003).

Large and novel disturbances are likely to increase as a consequence of climate change (Hobbs et al. 2006; Westerling et al. 2006). In interior

1

TABLE 1.1
Estimated mean annual timber losses (m³/yr) in Europe* from different natural disturbances

Agent of Disturbance	1950s	1960s	1970s	1980s	1990s	2000s
Fire	—	2,052,112	6,244,587	6,626,240	7,864,898	10,897,001
Wind	7,635,135	23,155,976	14,170,486	14,251,394	39,318,731	32,659,283
Snow	302,771	969,502	1,309,577	2,103,495	886,395	1,656,292
Other abiotic damage	212,617	76,785	320,706	6,495,915	2,395,349	—
Bark beetles	2,226,033	392,812	2,050,340	5,351,412	8,742,106	3,431,657
Fungal pathogens	—	—	—	1,284,609	1,824,344	—
Other biotic damage	—	363,360	643,452	4,398,105	5,923,307	—
Anthropogenic damage	—	217,503	425,548	1,681,173	1,009,898	—
Other	—	145,456	1,078,933	3,666,922	5,636,121	—
Total	—	27,373,506	26,243,628	45,859,266	73,601,149	—

Source: Schelhaas et al. (2001) updated to 2006 by M.-J. Schelhaas (personal communication).
* includes Includes Turkey, but not countries of the former USSR.
— incomplete data

British Columbia, more than 13 million hectares of lodgepole pine (*Pinus contorta* var. *latifolia*) forest have been affected by the mountain pine beetle (*Dendroctonus ponderosae*) (British Columbia Ministry of Forests and Range 2006, 2007) (fig. 1.1), and the epidemic of this native pest has spread into neighboring Alberta and could even invade stands of a new host species, jack pine (*Pinus banksiana*). Such unusual behavior of the mountain pine beetle is believed to be the result of unprecedented expanses of mature pine forest (due to widespread forest fires a century ago and subsequent fire control), coupled with warming winter temperatures and consequent improved survival of overwintering insect broods (Carroll et al. 2004).

Natural and Other Disturbances

Throughout this book we refer to natural disturbances. As is further described in chapter 2, these disruptions are recognized as discrete events in time and place that cause widespread mortality of trees. We distinguish such unplanned disturbances that "nature" generates from scheduled hu-

FIGURE 1.1. A landscape in interior British Columbia, Canada, which was recently dominated by lodgepole pine but has been dramatically altered by an epidemic of mountain pine beetle and subsequent salvage logging (photo by David Lindenmayer).

man disturbances of forests, such as timber harvesting, road building, and land clearing for agriculture. Although varying in importance around the world, many or most forest fires in a region may originate from human ignitions, and other disturbances such as landslides can also be induced by human actions. It is debated whether such forests would eventually burn from lightning ignitions anyway, and whether modern rates of human-caused ignitions are more or less than those practiced by indigenous peoples for millennia. Insect populations preying on forest trees may have been introduced from another continent, or may be undergoing exponential growth as a result of increasing temperatures or the loss of predators. The frequency and strength of many weather-related disturbances such as hurricanes seem to be a consequence of global warming induced by fossil fuel consumption, and are expected to continue increasing in the future (Dale et al. 2001). As such, it could be argued that a large proportion of the "natural disturbances" affecting forests have their origin in human activities.

Salvage Logging

Salvage logging is a common response to natural forest disturbances. Salvage logging is defined by the Society of American Foresters (Helms 1998, 158) as "the removal of dead trees or trees damaged or dying because of injurious agents . . . to recover economic value that would otherwise be lost."

In practice, salvage logging often also results in the removal of undamaged live trees along with the dead or damaged ones (Shore et al. 2003; Foster and Orwig 2006). Salvage logging is widely practiced throughout the world (table 1.2).

Salvage logging may occur after a wide variety of disturbances, such as the following:

- Wildfires (McIver and Starr 2000; Nappi et al. 2004; Hutto 2006)
- Insect attacks (Radeloff et al. 2000; Shore et al. 2003; Foster and Orwig 2006)
- Hurricanes, cyclones, and other severe windstorms (Nilsson 1975; Savill 1983; Foster and Orwig 2006)
- Floods (Gregory 1997) and debris flows (Beschta, 1979)
- Volcanic eruptions (Eggler 1948; Franklin et al. 1985; Dale et al. 2005)

Most of the literature on salvage logging is related to wildfires (table 1.2), although there are regional differences in the dominant disturbance agent: fires in Australia and western North America, windstorms in Europe and eastern North America (table 1.2), and insects and fire in Canada and, more generally, in boreal regions.

Salvage logging is practiced in all forested regions of the world including tropical forests (Van Nieuwstadt et al. 2001) and boreal and temperate forests (McHugh 1991; Che and Woen 1997; Kulakowski and Veblen 2007). It is also applied in plantations (Holtam 1971; Shakesby et al. 1993, 1996; Rackham 2001). Salvage logging is most prevalent in ecosystems where natural disturbances occur as partial or complete stand-replacement events (McIver and Starr 2000; Lindenmayer and Noss 2006).

Salvage logging has been practiced for a long time and is a concept and practice dating back to the origins of forestry. Enormous volumes of Norway spruce (*Picea abies*) were salvaged after being killed by insects in the mid-nineteenth century and again in the mid-twentieth century (Bejer 1988; Christiansen and Bakke 1988). Salvage logging of portions of the 1902 Yacolt Burn were among Weyerhaeuser Company's earliest operations in the U.S. Pacific Northwest. Forests in western Oregon damaged by the 1933 Tillamook Burn were salvaged continuously until at least 1959 and probably 1971 (Wells 1998). Wet eucalypt forests in Australia were salvaged after wildfires in 1926, 1932, and 1939 (Lindenmayer and Ough 2006; see figure 1.2 and chapter 4).

Salvage logging programs can be substantial and prolonged. One of the largest salvage programs in the United States followed the 1938 Hurricane in the New England region (Foster and Orwig 2006). Large areas of private, state, and federal timber affected by the 1980 Mount St. Helens eruption were salvage logged (Franklin et al. 1985, 1995). Harvesting of fire-damaged timber near Campbell River on Vancouver Island provided wood to mills in Vancouver, British Columbia, for at least a decade (Mackie 2000). After Storm Gudrun (called Erwin in Germany) struck Sweden in January 2005, forest-harvesting equipment and operators were mobilized from across northern Europe to rapidly salvage most of the 7.5 million cubic meters of damaged timber (Sondell 2006; fig. 1.3).

Salvaged timber now represents a significant percentage of the wood harvested in many regions of the world. At least 78 percent of the wood volume damaged by natural disturbances in Europe is salvaged (Schelhaas et al. 2001), generating 8 percent of the annual wood harvest (Schelhaas et al. 2003). In 2005, approximately 40 percent of logging on federal forestlands in the USA was focused on salvaging timber from fire-damaged stands. In the states of Washington and Oregon, the proportion of timber from salvage logging increased from 14 percent between 1980 and 1988 to 21 percent between 1989 and 1998 (McIver and Starr 2000). Salvage of dead tree boles now represents 20 percent of British Columbia's current cut from public forests.

Economic Rationale for Salvage Logging

Many justifications for salvage logging have been offered, but the most obvious and defensible rationale is economic (Cannell and Coutts 1988; Ulbricht et al. 1999; Akay et al. 2006). Salvage logging can capture some of the economic value in dead and damaged trees that would otherwise be lost, whether in the form of timber, pulpwood, or fuelwood (U.S.

FIGURE 1.2. Salvage logging in the Central Highlands of Victoria following the 1983 wildfires. A. Some trees survived the wildfire but were removed in subsequent salvage logging operations. B. Replanting efforts initially failed and the second attempt to reestablish tree cover took place. C. Ten years after salvage logging and replanting, stands with highly simplified structural complexity dominate the landscape (photos by David Lindenmayer).

FIGURE 1.3. One million cubic meters of salvaged timber stockpiled on an unused airstrip near Byholma, Sweden, following the 2005 Gudrun storm. Such stockpiles are kept wet with sprinkler systems to reduce deterioration by fungi and insects until the wood can be processed (Photo by Ulf-Göran Wannerfledt, courtesy of Vida AB).

Government Accountability Office 2006; Prestemon et al. 2006). Some consumers may perceive that wood products from salvaged dead trees are more environmentally friendly than products made from wood harvested from green stands (Donovan 2004).

When large amounts of timber are salvaged, that timber is typically substituted for green (live) wood that would have been harvested during that same time period. Large salvage logging programs may generate so much timber that they disrupt markets by flooding them with wood (Prestemon et al. 2006). This can significantly disrupt revenue streams (Kohnle and von Teuffel 2004; Luppold and Sendak 2004) and disrupt the future availability of resources for harvesting in the medium and long term (Eggler 1948; Spittlehouse and Stewart 2003). Large salvage logging programs also may generate wood in excess of regional processing (mill) capacity, necessitating creative responses from wood products industries. For example, a five-year supply of logs was salvaged and preserved in ponds and lakes until needed following the 1938 New England Hurricane. The 1962

TABLE 1.2
Documented examples of salvage logging following different kinds of natural disturbance events

Example	Location	Citation(s)
Wildfires		
1902 Yacholt Fire	Washington State, USA	Holbrook (1943)
1933 Sayward Fire	Vancouver Island, British Columbia, Canada	Mackie (2000)
1933 Tillamook Burn	Oregon, USA	Isaac and Meagher (1938), Wells (1998)
1938 Campbell River Fire	Vancouver Island, British Columbia, Canada	Mackie (2000)
1939 Black Friday Fires	Central Highlands of Victoria, Australia	Noble (1977), Lindenmayer and Ough (2006)
1970 Entiat River Fire, Wenatchee National Forest	Washington State, USA	Klock (1975), Helvey et al. (1985)
1972 Prescott National Forest	Arizona, USA	Blake (1983)
1977 Fire	Maine, USA	Hansen (1983)*
1977 Hog Fire	California, USA	Stuart et al. (1993)
1981 Fire	Nevada, USA	Johnson et al. (2005)
1983 Ash Wednesday Wildfires	Central Highlands of Victoria, Australia	Smith and Woodgate (1985), McHugh (1991)
1987 Klamath National Forest Fire	California, USA	Hanson and Stuart (2005)
1987 Silver Fire	Oregon, USA	Thompson et al. (2007)
1987 Hotelling Fire	California, USA	Stuart et al. (1993)
1988 Clover-Mist Fire	Wyoming, USA	Simon et al. (1994)
1989 Mt. Carmel Fire	Israel	Haim and Izhaki (1994), Inbar et al. (1997), Ne'eman et al. (1997)
1991 Valle Grande Fire, Carson National Forest	New Mexico, USA	Cram et al. (2006)
1991 Blackfoot-Clearwater Wildlife Management Area Fire	Montana, USA	Hutto and Gallo (2006)
1992 Foothills Fire	Idaho, USA	Saab and Dudley (1998), Russell et al. (2006)
1992 Eagle Fire	British Columbia, Canada	Khetmalas et al. (2002)
1994 Tyee Fire, Wenatchee National Forest	Washington State, USA	Carroll et al. (2000)
1994 Star Gluch Fire	Idaho, USA	Saab and Dudley (1998), Saab et al. (2007)
1994 Rat Creek Fire	Washington State, USA	Haggard and Gaines (2001)
1994 Tus Valley Fire	Albacete Province, Spain	Martínez-Sánchez et al. (1999)
1995 Hawk Fire	Saskatchewan, Canada	Morissette et al. (2002)

TABLE 1.2
Continued

Example	Location	Citation(s)
1996 Horseshoe Fire, Coconino National Forest	Arizona, USA	Cram et al. (2006)
1996 Summit Wildfire	Oregon, USA	McIver and McNeil (2006), McIver and Ottmar (2007)
1997 Val-Paradis Fire	Quebec, Canada	Nappi et al. (2003), Purdon et al. (2004)
1997–1998 rain-forest fires	East Kalimantan, Indonesia	Van Nieuwstadt et al. (2001)
1998 Chip Lake Burn	Alberta, Canada	Schmiegelow et al. (2006)
2000 Bitterroot National Forest Fire	Montana, USA	American Lands Alliance (2003), Prestemon et al. (2006)
2001 Chisholm Fire	Alberta, Canada	Phillips et al. (2006)
2001 Eldorado and Tahoe National Forest Fires	California, USA	Thrower (2005)
2002 Biscuit Burn	Oregon, USA	Donato et al. (2006a, 2006b), U.S. Government Accountability Office (2006), Thompson et al. (2007)
2002 Penasco Fire, Lincoln National Forest	New Mexico, USA	Cram et al. (2006)
2002 Hayman Fire	Colorado, USA	Robichaud et al. (2003)
2002 Rodeo-Chediski Fire	Arizona, USA	USDA Forest Service (2004)
2002 Missionary Ridge Fire	Colorado, USA	American Lands Alliance (2003)
2002 McNally Fire	California, USA	American Lands Alliance (2003)
2002 Toolbox Fire	Oregon, USA	American Lands Alliance (2003)
2002 House River Fire	Alberta, Canada	Bradbury (2006); Schmiegelow et al. (2006), Macdonald (2007)
2003/2004 Eastern Victorian Wildfires	Eastern Victoria, Australia	Victoria Dept. of Sustainability and Environment (2003)
2006/2007 Eastern Victorian Wildfires	Eastern Victoria, Australia	Victoria Dept. of Sustainability and Environment (2007)
1991 Águeda Basin Plantation fires	North-central Portugal	Shakesby et al. (1993, 1996)
Insect Outbreaks		
1853–1863. Nun moth and subsequent bark beetles, spruce forests	Western Russia and East Prussia (modern-day Poland)	Bejer (1988)

TABLE 1.2
Continued

Example	Location	Citation(s)
1970s. Spruce budworm, fir and spruce forests	Maine, USA; New Brunswick and Quebec, Canada	Sader et al. (2005), Etheridge et al. (2006)
Early 1980s. Gypsy moth outbreak in mixed oak forests	New England states and Pennsylvania, USA	Sewell and Brown (1995), Schuler et al. (2005)
1988–ongoing. Hemlock woolly adelgid, northern hardwood forests	Connecticut and Massachusetts, USA	Orwig and Foster (1998), Brooks (2001), Foster and Orwig (2006)
1990–1995. Jackpine budworm, Pine Barrens region	Wisconsin, USA	Radeloff et al. (2000)
1995–ongoing. Mountain pine beetle, interior montane and subboreal forest	British Columbia, Canada	Hughes and Drever (2001); Shore et al. (2003); Safranyik and Wilson (2006)

Windstorm, Cyclone, Hurricane

1938 hurricane	New England states, USA	Peart et al. (1992), Foster et al. (1997), Cooper-Ellis et al. (1999), Foster and Orwig (2006)
1949 windstorm	Wisconsin and other Upper Midwest states, USA	Stearns and Likens (2002)
1950 blowdown, Adirondack Mountains	New York State, USA	Dobson et al. (1990)
1962 Columbus Day Windstorm	Oregon and Washington State, USA	Lynott and Cramer (1966)
1968 Scottish windstorm	Scotland	Holtam (1971)
1969 Den Stora Stormfällning	Sweden	Morling 1981 in Dobson et al. (1990)
1973 and 1983 windstorms, Bull Run Basin	Western Oregon, USA	Sinton et al. (2000)
1974 Canberra Windstorm	Australian Capital Territory, Australia	Cremer et al. (1977)
1987 British hurricane	United Kingdom	Cannell and Coutts (1988)
1989 Hurricane Hugo	South Carolina, USA	Marsinko et al. (1993), Fail (1999), Amatya et al. (2006), Prestemon and Holmes (2004)
1990 storm Vivian	Switzerland	Angst and Volz (2002)
1995 Adirondack Storm[†]	New York State, USA	Robinson and Zappieri (1999)
1995 Hurricane Opal	North Carolina, USA	Elliott et al. (2002), Greenberg (2002)
1997 Routt National Forest Blowdown	Colorado, USA	Kulakowski and Veblen (2007)

TABLE 1.2
Continued

Example	Location	Citation(s)
1999 Windstorm	Tennessee, USA	Leach and Peterson (unpublished data)
2005 Hurricanes Katrina and Rita	Louisiana, Alabama, Mississippi, Texas, USA	Stanturf et al. (2007)
2005 Hurricane Gudrun	Sweden	Sondell (2006)
2006 Bear Tornado, Payette National Forest	Idaho, USA	USDA Forest Service (2006)
Other Kinds of Disturbance		
1900s Chestnut Blight	Eastern USA	Frothingham (1924)
1965/1966 Debris flows	Oregon, USA	Beschta (1979)
1980 Mount St. Helens volcanic eruption	Washington State, USA	Franklin and MacMahon (2000), Crisafulli et al. (2005); Titus and Householder (2007)
1998 Ice storm	Quebec and Ontario, Canada; New York, Vermont, New Hampshire, USA	Irland (1998), Millward and Kraft (2004)
2004 Boxing Day Tsunami	Various coastal island and mainland locations around the Indian Ocean	Lindenmayer and Tambiah (2005)

*The 1977 wildfire was preceded by a severe windstorm and windthrow in 1974.
†Salvage logging was initially proposed, but plans were later aborted (Robinson and Zappieri 1999).

Columbus Day windstorm in the U.S. Pacific Northwest blew down 2,360,000 cubic meters of timber, most of which was salvaged; the wood products industry created an Asian log export market to move this large volume of salvaged timber.

Other Rationales for Salvage Logging

Many noneconomic justifications have also been proposed for salvage logging. There is, of course, the natural tendency of societies to want to bring order back from the "chaos" created by large disturbance events (Hull 2006). In particular, Western societies find the perception of waste and disorder to be anathema (Noss and Lindenmayer 2006). Hence, major efforts may be mounted to "clean up" after large, intense disturbance events.

The following are some of the specific justifications offered for salvage beyond economics or general societal inclinations to restore order:

- Safety concerns, such as the hazards created by large numbers of standing dead trees or snags for people working or recreating in burned areas (Ne'eman et al. 1997; Shore et al. 2003).
- Reducing fuels available for subsequent fires, on the basis that the dead wood provides abundant fuels and increases fire risk for re-burns on the disturbed area as well as the adjacent forest (Sessions et al. 2004; British Columbia Ministry of Forests and Range 2006, 2007; Passovoy and Fule 2006). Snags are a particular concern as they can generate firebrands (Inbar et al. 1997; Orwig and Kittredge 2005).
- Reducing fuels to limit the amount of smoke produced in subse-quent prescribed burns and, in turn, limit the effects of smoke on nearby settlements (Achtemeier 2001).
- Reducing the potential for dead and dying trees to become breeding grounds for pests and pathogens that would invade adjacent unaf-fected forests (Holtam 1971; Christiansen and Bakke 1988; Amman and Ryan 1991; Hughes and Drever 2001). For example, outbreaks of spruce beetle (*Dendroctonus rufipennis*) in the U.S. state of Col-orado have traditionally followed extensive windthrow events (Schmidt and Frye 1977; Lindemann and Baker 2001).
- An expectation that some damaged trees will inevitably die from pest infestations, and the desire to remove them before they are killed (Brooks 2004; Foster and Orwig 2006).

Salvage logging has also been justified on the basis that it will con-tribute to ecological recovery of naturally disturbed forests (Sessions et al. 2004). This is, in part, based on the following perceptions:

- Legacies of dead and damaged trees do not contribute to the recov-ery of ecosystem processes and biodiversity (Sessions et al. 2004).
- Ecosystem recovery will be assisted or accelerated by speeding the re-establishment of forest cover (Bartlett et al. 2005; Sessions et al. 2004).
- Naturally disturbed areas have limited value for biota (Morissette et al. 2002).

The notion that salvage logging assists the ecological recovery of natu-rally disturbed forests is fundamentally incorrect (Lindenmayer et al. 2004). Hence, justifications for salvage logging based on contributions to ecological recovery have little merit. We know of few circumstances where salvage logging has been demonstrated to directly contribute to recovery of ecological processes or biodiversity. Under some circumstances, salvage logging can indirectly contribute through generation of funds for restorative activities. Conversely, there is abundant theoretical and empirical evidence (presented later in this volume) that salvage logging interferes with natural ecological recovery (Lindenmayer and Ough 2006) and may increase the likelihood and/or intensity of subsequent fires (e.g., Kulakowski and Veblen 2007; Thompson et al. 2007). As Jerry Franklin stated,

> based on our current understanding of forest recovery following disturbances, timber salvage is most appropriately viewed as a "tax" on ecological recovery. The tax can either be very large or relatively small depending upon the amount of material removed and the logging techniques that are used (2005, 2).

Indeed, the term *salvage* itself is problematic ecologically given that the word means "recovering or saving" (Lindenmayer and Noss 2006). We return to this issue of terminology in chapter 6.

Importance of This Book

We based this book on the following premises:

• Large forest disturbances have been and will remain significant and prominent events, with the possibility that there will be increasing numbers and novel forms of disturbances as a consequence of climate change.
• Society has come to consciously value forests much more broadly than simply as sources of wood, such as for the ecological services that they provide (e.g., watershed protection) and for the biodiversity that they support.
• There is much relevant scientific literature that should be considered during the development of policies and management approaches regarding disturbed forest landscapes.

Our objective in this book is to examine the *ecological* consequences of salvage logging for a broad audience of stakeholders. We recognize that ecological values are only one of the criteria society uses in making forest management decisions, including post-disturbance activities such as salvage logging. Similarly, science provides only one of the relevant knowledge bases. Nevertheless, we believe that a synthesis of ecological science relevant to salvage logging is important to fully inform public dialogue. When decisions are made regarding salvage and other post-disturbance activities, they should be made with clarity about the rationale for those activities.

A detailed consideration of salvage logging appears particularly timely in view of climate change. There is increasing evidence that one of the results of climate change is that large disturbance events are becoming more frequent (Cary 2002), widespread (Flannigan et al. 2005; British Columbia Ministry of Forests and Range 2006), intense (Emanuel 2005), or all of these (Franklin et al. 1991; Lenihan et al. 2003; Pittock 2005). There is already statistical evidence that as a result of climatic change there has been an increase in the extent and intensity of wildfires (Westerling et al. 2006), more numerous and intense hurricanes (Goldenberg et al. 2001), and a massive collapse of tree populations under novel infestations of native insects (Breshears et al. 2005). Hence, there is the potential for salvage logging proposals to become even more widespread and frequent (Schelhaas et al. 2003; Spittlehouse and Stewart 2003).

Although numerous volumes deal with ecological impacts of traditional ("green") logging and ways they can be mitigated (e.g., DeGraaf and Miller 1996; Franklin et al. 1997; Hunter 1999; Burton et al. 2003b; Lunney 2004), no equivalent volume deals with salvage logging. This book is an attempt to redress this imbalance.

Finally, we acknowledge the primacy of management objectives in determining appropriate post-disturbance management activities for affected areas, including the relative weights given to ecological and economic values. The process of evaluating various courses of action should begin with a consideration of management goals for the property, whether they involve wood production, watershed protection, conservation of biodiversity, or, as is most often the case, a mixture of economic and ecological goals.

Structure of This Book

This book is composed of a series of short, linked chapters. First, chapter 2 presents a brief overview of natural disturbance regimes in forests as a basis for examining issues associated with salvage logging. Potential ecological impacts of salvage logging are considered in chapter 3. Brief reviews of a wide variety of salvage logging projects are provided in chapter 4; these short case studies aim to highlight the diversity of issues associated with salvage logging. Chapter 5 is a discussion of the importance of management objectives in determining post-disturbance management policies, since the spatial scale and intensity of salvage logging should relate to those objectives. Chapter 6 also considers some forest management policy implications of salvage logging. A glossary that defines key terms associated with salvage logging completes this book.

Limitations of This Book

Our focus is on forests because this is where salvage logging is most commonly conducted and where it is most controversial. However, we do acknowledge that salvaging of damaged trees can occur in other kinds of environments, such as city parks and boulevards, intensively managed tree plantations or tree farms, aquatic ecosystems following floods (Gregory 1997; Mitchell 2007; Burley et al. 2008), and on beaches, including those affected by tsunamis (Bryant 2003; Lindenmayer and Tambiah 2005). Salvaging of natural resources other than trees may occur under a variety of circumstances—such as harvesting of storm-damaged fruit or grain crops—but these are also not considered in this book.

This book focuses on the ecological effects of salvage logging. We give little consideration to many of the other possible post-disturbance management activities, such as grass seeding and tree planting. The effectiveness of many of these activities in achieving stated goals, let alone consideration of their ecological impacts, needs serious objective assessment by policymakers and managers (e.g., see Beyers 2004). Similarly, the topic of ecological effects of management actions prior to or during disturbances (e.g., fire suppression activities, such as back-burning [blackout burning]) is given minimal consideration, although the potential for significant positive and negative impacts is large (Backer et al. 2004).

This book is weighted toward consideration of postfire conditions and management because of the overwhelming dominance of this disturbance type in the existing literature. Management after wildfire, including salvage logging, has received far more attention than other kinds of natural disturbances (see table 1.2). Nevertheless, we attempted to include as much material as possible relevant to salvage logging after windstorms and other natural disturbances.

We deliberately kept this book brief as we are cognizant that the present age of the "information superhighway" has, for many, become the era of the "information superglut." We suspect that few, if any, scientists, managers, policymakers, or stakeholders have been able to keep up with an exponentially expanding literature. Hence, we were brief in the hope that interested parties who need to know about natural disturbances and recovery and the role of salvage logging will find time to access the material summarized in this volume.

Some topics are probably not covered in the detail that some people will think is warranted; comprehensiveness is the trade-off in our effort at brevity. For example, research on natural disturbance in forests (chapter 2) is a massive arena, a comprehensive summarization of which probably exceeds the capacity of any individual or small group of people. Similarly, the body of research on impacts of traditional (i.e., nonsalvage) logging is also vast and, consequently, can be dealt with in only a cursory fashion in a short book. We apologize to colleagues who might think that we have short-changed their work. This was not intentional. Rather, it is a result of our pragmatic approach of trying to give readers at least a flavor of the key issues associated with ecological effects of salvage logging rather than to comprehensively review and cite all literature relevant to this topic.

Chapter 2

Natural Disturbance of Forest Ecosystems

Natural disturbances are increasingly recognized by ecologists as critically important ecosystem processes that help create habitats and resources for biological diversity (Connell 1978; Sousa 1984; Parr and Andersen 2006). Many aspects of composition and structure in forests at the tree, stand, ecosystem, and landscape scales are shaped by natural disturbances (Parminter 1998; Cary et al. 2003; Frelich 2005). In addition, many species have strong associations with natural disturbance regimes—some positive and some negative (Bergeron et al. 1993; Bunnell 1995; Spies et al. 2004).

Salvage logging, by definition (see chapter 1), follows large-scale and/or high-intensity natural disturbance. Therefore, any discussion of salvage impacts must first be preceded by a discussion of natural disturbance. This chapter outlines the primary features of natural disturbance regimes. This includes a short commentary on one of the most critical aspects of natural disturbances: the biological legacies that they leave behind and how these legacies influence ecosystem processes, biodiversity responses, and ecological recovery (Hansen et al. 1991; Franklin et al. 2000). We also explore the importance of using natural disturbances as a template to guide human disturbance regimes. This topic is extensively discussed in the forest management literature (e.g., Hunter 1993, 2007; Mitchell et al. 2002; Buddle et al. 2006), but it needs to be reemphasized as a basis for subsequent discussions of salvage logging.

Some Important Features of Natural Disturbance Regimes

Natural disturbances vary spatially and temporally in forests. They range from frequent, low-intensity, gap-forming disturbances operating at the scale of individual trees (Bray 1956; Runkle 1981; Denslow 1987; Lertzman 1992; Yamamoto 1992) to infrequent, landscape-scale, high-intensity events (fires, windstorms, insect attacks) that can radically alter stands and sets of stands and landscapes (Turner et al. 1997; Spies and Turner 1999; Stocks et al. 2002). In addition, a region may experience unprecedented disturbances, such as glaciation or the introduction of exotic herbivores or pathogens. This book is largely concerned with salvage logging following infrequent, large-scale, high-intensity disturbance events. This broad category of natural disturbances can be defined as discrete events that do not have a substantial human origin and which alter ecosystem structure and resource availability (after White and Pickett 1985).

Large-scale and/or high-intensity natural disturbances typically result in a rapid release or reallocation of community resources (Shiel and Burslem 2003). They are characteristic of many terrestrial ecosystems (Pickett and Thompson 1978; Agee 1993; fig. 2.1) and can include the following events:

FIGURE 2.1. Aftermath of major disturbances in forests: A. Volcanic eruption (photo by Jerry Franklin). B. Wildfire (photo by Phil Burton). C. Insect attack (photo by Phil Burton; inset western hemlock looper, *Lamdina fiscellaria lugubrosa*, photo by Bob Duncan, courtesy of Canadian Forest Service). D. Windstorm (photo by David Lindenmayer).

FIGURE 2.1. (Continued)

- Wildfires (Luke and McArthur 1977; Wadleigh and Jenkins 1996; Gill et al. 1999)
- Windstorms, hurricanes, and cyclones (Foster and Boose 1992; Peterson and Pickett 1995; Schnitzler and Borlea 1998; Ulanova, 2000; Schoener et al. 2004)
- Ice storms (Dugay et al. 2001; Tremblay et al. 2005)
- Volcanic eruptions (Franklin 1990; Leathwick and Mitchell 1992; Lavigne and Gunnell 2006)
- Floods (Swanson et al. 1998; Yarie et al. 1998; Calhoun 1999)
- Landslides (Ogden et al. 1996; Veblen et al. 1996; Geertsema and Pojar 2007)
- Widespread insect attacks (Holling 1992; Cogbill 1996; Shore et al. 2003)

BOX 2.1.
Natural Disturbance in Canadian Forests

The forests of Canada experience some of the largest natural disturbances in the world. Far more forest is affected by natural disturbance than human disturbance (see figure 2.2). Insect attacks by species such as the spruce budworm caterpillar (*Choristoneura fumiferana*) regularly affect large areas of boreal forest. The current outbreak of the mountain pine beetle is killing millions of hectares of lodgepole pine in western Canada (see chapter 4 for a detailed case study). Wildfire is another important mechanism of disturbance in Canadian forests (Rowe and Scotter 1973; Burton et al. 2003b, 2007; fig. 2.2), and most of the flora and fauna is well adapted to it. Fires are fueled by thick organic layers that accumulate under the cold, damp conditions of northern forests and peatlands, but which regularly dry out and become very flammable. Some species of conifers are especially flammable, as is the case of black spruce (*Picea mariana*), which often has a crown that extends to the forest floor. The amount of forest and woodland burned in Canada, including the boreal region, has been increasing in the past few decades (Stocks et al. 2002). Approximately 10 percent of Manitoba's forest and woodland burned in 1989 alone, with another 4 percent burning in 1994. A staggering 17.5 percent of the forest and woodland in the Northwest Territories burned in 1994 and 1995 combined (Canadian Ccouncil of Forest Ministers 2007).

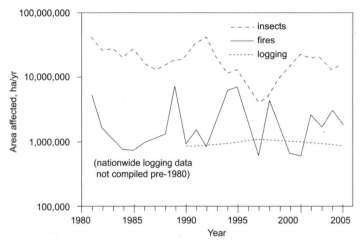

FIGURE 2.2. Canadian forest area affected by different natural disturbances from 1981 to 2005. Because logging includes some salvage of trees killed by fires and insects, there is some double counting in the areas tallied by these statistics. Although forest affected by insect outbreaks rarely suffers complete tree mortality, it is clear that insect outbreaks are the most widespread form of forest disturbance in Canada (data from Canadian Council of Forest Ministers 2007).

• Prolonged droughts (Gordon et al. 1988)
• Earthquakes
• Tsunamis (Bryant 2001; Sharma 2005)

Natural disturbance regimes are inherently variable, and the local intensity of most disturbances is dictated by a large degree of historical contingency and chance (Kulakowski and Veblen 2007). Hence, no two natural disturbances will be identical (Johnson et al. 2005; Schoener et al. 2004). This variability also produces multiple (and often simultaneously acting) post-disturbance pathways (Noble and Slatyer 1980; Turner et al. 1998; Lecomte et al. 2006). This is illustrated in many mesic forest types such as those found in the U.S. Pacific Northwest (Halpern 1988; Morrison and Swanson 1990), the boreal forests of Canada (Burton et al. 2003b), and the wet eucalypt forests of southeastern Australia (Mackey et al. 2002).

Large-scale and/or high-intensity natural disturbances vary substantially in many ways:

- Timing (e.g., the time of the year or time of day when they occur)
- Frequency (or its inverse, return interval)
- Intensity
- Size
- Heterogeneity (i.e., variation in intensity and impact within the limits of the total area affected)
- Duration (e.g., fires and windstorms can be relatively short lived, whereas drought and insect attack can be prolonged)
- Biological legacies (see below)

These factors act together rather than in isolation. For example, fire frequency and fire intensity usually co-vary. Many landscapes are characterized by recurrent low-intensity disturbances, but few landscapes naturally experience frequent high-intensity wildfires. Factors such as climatic conditions and topography further influence how variables interact within particular landscapes (Rülcker et al. 1994; Lindenmayer et al. 1999c; Mackey et al. 2002). It is the particular combination of cause (agent), size, frequency, and severity of disturbances that prevail in a landscape that characterizes "disturbance regimes" (Gill 1975; Swanson et al. 1994).

Variation in disturbance regimes leads to marked differences in landscape and stand conditions by significantly influencing the number, type, and spatial distribution of habitat patches (e.g., age cohorts of stands) as well as stand-level biological legacies (Franklin et al. 2000; Turner et al. 2003). Such temporal and spatial variability in natural disturbance regimes helps explain regional differences in vegetation patterns and species assemblages (Boose et al. 2004). For example, areas of western Canada subject to high natural fire frequencies support more taxa typical of early successional forests than do regions with longer disturbance return intervals (Bunnell 1995).

Non-uniform patterns of organism distribution and abundance result (in part) from spatial variation in environmental conditions, such as climate, terrain, soils, and availability of moisture and nutrients (e.g., Woodward and Williams 1987; Prentice et al. 1992; Hansen and Rotella 1999). Such spatial variation leads to natural ecosystems being heterogeneous at the landscape scale (Huggett and Cheesman 2002; Parr and Andersen 2006). Disturbance regimes overlay and interact with the patterns created by environmental regimes to further shape heterogeneous landscapes and influence the distribution of species (Wardell-Johnson and Horowitz 1996; Boose et al. 2004). Factors such as topographic variability and weather changes during a disturbance event can result in undisturbed patches within the broad boundaries of a disturbed area, such as a fire or windstorm (Eberhart and Woodard 1987; Syrjänen et al. 1994). Indeed, recent research has revealed considerable internal variability in the severity and residual structure of disturbance events typically mapped as a single change in land cover (e.g., Kafka et al. 2001; Schmiegelow et al. 2006; Burton et al. 2008).

It is a challenging task to characterize the historical disturbance regime of forest stands and landscapes, but this is essential to understand how site, succession, and disturbance interact to produce the patterns we see on the land today. Such work may be based on the dendrochronological dating of tree ring scars caused by fires, treefalls, avalanches, and insect outbreaks (e.g., Frelich and Lorimer 1991; Veblen et al. 1994), or the mapping of historically observed disturbance events (e.g., Morgan et al. 2001; Schulte and Mladenoff 2005). In most cases, there is an attempt to characterize the causes and levels of tree mortality (and sometimes other measures of severity or ecosystem impact), their spatial extent, and their frequency over time. The many combinations of disturbance agents (and their severity, homogeneity, extent, and frequency) combine with underlying differences in terrain and climate to generate distinctive stand development trajectories (Frelich 2005). It is the result of these interactions that we then recognize as distinctive forest types, varying in age, composition, and structure.

There are many good reasons for emulating key features of the historical natural disturbance regime in managed landscapes (Landres et al. 1999; Bergeron et al. 2002; Perera et al. 2004). But researchers consistently find that disturbance regimes have changed in the past as a result of climate fluctuations and shifts in human influence (e.g., Weir et al. 2000; Wallace 2004), and they widely expect these regimes (especially fire regimes) to be altered by climate changes in the future (Dale et al. 2001; Flannigan et al. 2005). Consequently, there is a need to understand and predict disturbance regimes on the basis of underlying causal factors, and how those might change in the future (e.g., Quine et al. 2002; Cleland et al. 2004; Taylor et al. 2005).

Large-scale Natural Disturbances, Key Ecosystem Processes, and Biotic Responses

Natural disturbances are key processes in the vast majority of ecosystems (Pahl-Wostl 1995; Frelich 2005). Relatively recent paradigms in ecology emphasize the dynamic, nonequilibrial nature of ecological systems of which disturbance is a recurring feature (Pickett et al. 1992; Holling et al. 1995; Cumming et al. 1996; Reice 2001). Complex interactions between relatively static and potentially equilibrial site features (including climate, soils, topography, flora and fauna) and dynamic, often nonequilibrial natural disturbance regimes, contribute to the maintenance of much of the world's biodiversity (e.g., Hansen and Rotella 1999; Shiel and Burslem 2003). Disturbances are likewise an important requirement for forest regeneration and the maintenance of primary productivity in many forest types. For example, important relationships exist between windstorms, treefall, soil fertility, and tree regeneration in the rimu (*Dacrydium cupressuim*) forests of southern New Zealand. Soil development in these forests is extremely slow. Norton (1989) estimated that the soil turnover half-life is about three thousand years. However, treefalls can boost soil fertility by bringing less-weathered (and more fertile) soil to the surface. Root mounds and fallen logs then provide places for trees and other plants to germinate (Duncan 1993).

An increasing body of evidence indicates that many species and ecosystems have evolved with, are adapted to, and depend on various types of natural disturbance (Bradstock et al. 2002; Burrows and Wardell-Johnson 2003; box 2.2). Indeed, there are well-developed ecological theories that link disturbance regimes and attributes of ecosystems such as the aggregate species richness (Brawn et al. 2001). One of these is the intermediate disturbance hypothesis, which forecasts highest species diversity at intermediate rates and intensities of disturbance (Connell 1978; fig. 2.3). The concept was developed to explain species richness on coral reefs and in rain forests (e.g., Rogers 1993; Aronson and Precht 1995; Molino and Sabatier 2001). It also has been applied to systems such as temperate forests and prairies (Collins et al. 1995). Relatively few species can survive frequent high-intensity disturbances, and most seem likely to be better adapted to either frequent low-intensity or infrequent high-intensity perturbation. In the absence of disturbance, sites can become dominated by a few competitive species or, for example, those tree species able to repro-

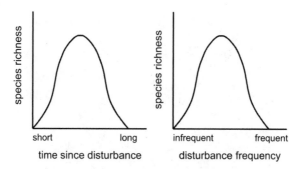

FIGURE 2.3. Relationships between species diversity and disturbance regimes under the intermediate disturbance hypothesis (redrawn from Wilson 1994).

duce in their own shade (Attiwill 1994a, 1994b). In contrast, ecosystems subject to intermediate levels of disturbance tend to have a mixture of pioneer and climax species. While empirical support for the intermediate disturbance hypothesis is inconsistent (e.g., Schwilk et al. 1997; Bascompte and Rodriguez 2000; Beckage and Stout 2000), it (and other disturbance-related theories) highlights the important range of roles that disturbances play in ecosystems.

The above discussion indicates that the vast majority of ecosystems (and the biota in them) have well-developed strategies for recovering from natural disturbance (e.g., Turner et al. 2003; fig. 2.4 and fig. 2.5). The existence of large forested landscapes over thousands and even millions of years in Asia, North America, South America, Africa, and Australia provides clear and unequivocal evidence that ecosystems can and do recover in the absence of human intervention. Good empirical data exist on natural recovery processes in intensively disturbed ecosystems, for instance, the response of the Greater Yellowstone ecosystem following the 1988 conflagration (Wallace 2004) and the Mount St. Helens ecosystem following the 1980 volcanic eruption (Franklin and MacMahon 2000; Dale et al. 2005). We further discuss these and other examples in chapter 4.

Natural Early Successional Habitats

The following are among the factors that make recently disturbed forests biologically diverse:

FIGURE 2.4. Examples of early post-natural disturbance recovery: A. Mount St. Helens blast zone, Washington, USA (photo by Jerry Franklin). B. Advance re-generation of balsam fir (*Abies balsamea*) released after spruce budworm (*Choristoneura fumiferana*) attack in Quebec, Canada (photo by Dan Kneeshaw). C. Natural regeneration by lodgepole pine (*Pinus contorta*) and Douglas-fir (*Pseudotsuga menziesii*) after forest fire in Jasper National Park, Alberta, Canada (photo by Phil Burton).

- A combination of surviving and pioneering species (Shiel and Burslem 2003)
- Diverse plant life-forms and structures (Stuart et al. 1993), which provide habitat for many organisms
- High availability of light and moisture
- A variety of microclimates (Society for Conservation Biology Scientific Panel on Fire in Western U.S. Forests 2005)

Some species, such as those shown in figure 2.5, are found only in, or are closely associated with, early successional environments . Many species of animals and plants (some rare or threatened) are attracted to or readily germinate in places that are burning or recently burned (Brawn et al. 2001; Imbeau et al. 2001; Fraser et al. 2004; Smucker et al. 2005). Taxa

A   B

C

FIGURE 2.5. Examples of species closely associated with early successional environments. A. Eastern chestnut mouse (*Pseudomys gracilicaudatus*, photo by Tony Robinson/Viridans Images). B. Mountain bluebird (*Sialia currucoides*, photo by Richard Hutto). C. Jewel beetle (*Castiarina sp.*, photo by Peter Valentine, James Cook University).

associated with early successional habitats include species dependent on charred standing trees and logs created by fires:

- Woodpeckers (Hutto 1995, 2006; Murphy and Lehnhausen 1998; Chambers and Mast 2005). For example, the black-backed woodpecker (*Picoides articus*) is often strongly associated with burned forests within two years following wildfire (Saab and Dudley 1998; Hoyt and Hannon 2002).
- Carnivorous mammals (Bull et al. 2001)
- Highly specialized beetles (Grove et al. 2002; Saint-Germain et al. 2004), such as those belonging to the families Buprestidae and Ceramycidae
- Bryophytes (Scott 1985; Pharo et al. 2004; Burton et al. 2006)

Early successional habitats within forests regenerating after windstorms can also be important for a range of biota. Studies in the United Kingdom have shown that open areas created by windthrow are valuable habitats for bryophytes (Quine et al. 2002) and the goshawk (*Accipter gentiles*) (Petty 1996). In Tierra del Fuego, South America, areas of densely packed windthrown trees limit browsing pressure of *Nothofagus* spp. seedlings by both native herbivores (the guanaco, *Lama guanicoe*) and domestic livestock, allowing vigorous tree regrowth to develop (JFF and DBL, personal observation). Turner et al. (2003) reported similar observations of fallen trees providing browsing-free zones in early successional aspen (*Populus tremuloides*) forests regenerating after the 1988 Yellowstone fires.

The importance of natural disturbance for the maintenance of ecosystem processes and biodiversity becomes apparent when humans alter natural disturbance regimes (box 2.2). Numerous examples exist. In many areas of ponderosa pine (*Pinus ponderosa*) forest in the USA (fig. 2.6), past fire suppression together with livestock grazing, old-growth logging, road building, and other human activities have created a forest condition highly susceptible to uncharacteristic stand-replacing fire, drought, and insect attack (Friederici 2003; Moore et al. 2004; Donovan and Brown 2007). The absence of fire in the mid-boreal forests of Sweden has created problems such as the lack of regeneration of particular plant species (Zackrisson 1977) and the loss of pyrophyllic species (Heliövaara and Väisänen 1984; Wikars 2001). Similarly, the absence of fire in far north

Queensland is leading to the loss of wet sclerophyll forest because it is being invaded by fire-sensitive rain-forest stands. This, in turn, threatens an array of species closely associated with wet sclerophyll forest and found nowhere else (Harrington and Sanderson 1994). The careful reintroduction of fire, in combination with selective tree removal, is now used as a forest restoration treatment in many parts of the world (Kuuluvainen et al. 2004; Brown et al. 2004; Douglas and Burton 2004; box 2.2).

Natural Disturbances and Biological Legacies

Natural disturbances in forests leave significant features of the original stand in the form of a rich array of biological legacies (Franklin et al. 1985; Hansen et al. 1991). Franklin et al. (2000, 9) define biological legacies as "organisms, organically-derived structures, and organically-produced patterns that survive from the pre-disturbance system."

Biological legacies represent the essential ingredients for secondary succession, and for distinguishing secondary succession from the much slower and sporadic primary succession on novel inorganic substrates—a feature recognized by early ecologists (Clements 1916; Drury and Nesbit

FIGURE 2.6. Fire restoration treatment in a ponderosa pine (*Pinus ponderosa*) forest (photo by Reed Noss).

BOX 2.2

Fire as a Key Ecosystem Process in Longleaf Pine Ecosystems

Fire is the primary natural disturbance in longleaf pine (*Pinus palustris*) stands in the southeastern United States (fig. 2.7a). Without fire, longleaf pine is replaced by hardwood trees (Conner and Rudolph 1989). In the past, continuous stands of longleaf pine were burned approximately every one to three years by fires ignited by lightning strikes that spread rapidly among adjacent and highly connected stands. In particular, the ground cover of old-growth stands facilitated the spread of fires. However, extensive landscape alteration means that remaining stands of longleaf pine are now spatially separated and the ground cover of regrowth stands does not carry fire as well as old growth. A crucial ecological process in longleaf pine — regular natural disturbances by fire — has been disrupted. This has had a significant negative impact on some important elements of the biota such as the endangered red-cockaded woodpecker (*Picoides borealis*; fig. 2.8b). This species constructs cavities in old, dying longleaf pine trees, but such trees have become rare as only remnant patches of large trees remain since the continuous expanse of 25 million hectares of primary forest has been reduced to about 4 million hectares of secondary forest (Tebo 1985; Simberloff 2004).

The management of longleaf pine ecosystems will require not only reestablishing fire but also engaging in targeted single-species work, such as translocations and the construction of artificial cavities for taxa of concern like the red-cockaded woodpecker (James et al. 1997; Conner et al. 2001; James et al. 2003). Notably, the natural burning regime in longleaf pine stands is frequent, low-intensity fire. This contrasts with infrequent, high-intensity, stand-replacing disturbances widely discussed in this book.

1973; McIntosh 1985). Biological legacies often constitute the nucleation points from which severely disturbed ecosystems start to recover, and which then facilitate subsequent growth in productivity and diversity (Connell and Slatyer 1977; del Moral and Bliss 1993; Turner et al. 2003). Small residual populations of animals also might serve to facilitate population expansion following natural disturbance (Whelan et al. 2002; Lindenmayer et al. 2005).

At the stand-level, biological legacies can include:

FIGURE 2.7. A. Longleaf pine (*Pinus palustris*) ecosystem (photo by Jerry Franklin). B. Red-cockaded woodpecker (*Picoides borealis*, photo by Reed Noss).

- Large living and dead trees (Fischer and McClelland 1983; Gibbons and Lindenmayer 2002; Fenger et al. 2006)
- Large logs (Harmon et al. 1986; Duncan 1993; Lindenmayer et al. 2002)
- Intact thickets of understory vegetation (Mueck et al. 1996; Ough 2002)

- Other vascular vegetation surviving in the form of seeds, rhizomes, and rootstocks (Whelan 1995; Lindenmayer and Ough 2006)
- Well-developed topsoils and duff layers (Johnson 1992; Haeussler and Bergeron 2004)
- Residual populations of animals (Whelan et al. 2002; Platt and Connell 2003)

As wildfires are heterogeneous in their intensity and impact across forest landscapes (Kafka et al. 2001; Shatford et al. 2007), they leave behind landscape-level biological legacies. These include unburned patches of forest (DeLong and Kessler 2000; Kotliar et al. 2007) and stands of forest burned at higher or lower intensities (Mackey et al. 2002; Thompson et al. 2007). Consequently, disturbance impacts, the types and densities of biological legacies, and the resulting patterns of recovery represent a nested complexity across a hierarchy of spatial scales.

Many factors influence the numbers, types, and spatial arrangements of biological legacies in stands and landscapes (Franklin et al. 2000):

- The type and intensity of disturbance (Smith and Woodgate 1985; Smucker et al. 2005)
- The time since the previous disturbance (Kulakowski and Veblen 2007), how intense it was, and how well developed that post-disturbance forest was. For example, the number and type of legacies in a burned or windblown old-growth stand will contrast strongly with those left in a naturally disturbed stand of regrowth forest.
- Additional stresses (e.g., drought, herbivory) before, during, and after the disturbance (Wisdom et al. 2006)
- The topography of the landscape (Lindenmayer et al. 1999c; Huggett and Cheesman 2002)
- Site characteristics such as elevation, soil moisture, and type of substrate (Broncano et al. 2005; Larson and Franklin 2005)
- Recovery strategies of organisms (e.g., resprouters, obligate seeders, etc., in the case of plants, as reviewed by Noble and Slatyer 1980; Whelan 1995)

Biological legacies at the stand and landscape levels significantly influence the rate and pathway of recovery of a post-disturbance ecosystem

FIGURE 2.8. Biological legacies following major natural disturbances. A. Advance regeneration of spruce (*Picea spp.*) and fir (*Abies lasiocarpa*) in a lodgepole pine (*Pinus contorta*) stand undergoing an insect attack in British Columbia, Canada, (photo by Phil Burton). B. Epicormic sprouting from standing boles in a eucalypt forest in southeastern Australia following wildfire (photo by David Lindenmayer). C. Elk (*Cervus canadensis*) in recently burned forest at Yellowstone National Park, Wyoming (photo courtesy of U.S. National Park Service).

(Franklin et al. 2000; Lindenmayer and Franklin 2002). They have many critical ecological roles:

- Surviving, persisting, and regenerating after disturbance and being incorporated as part of the recovering stand (e.g., producing forests with multiple cohorts [Lindenmayer et al. 1999c])
- Facilitating survival of other species in disturbed stands (Kavanagh and Turner 1994; Kotliar et al. 2007)
- Providing habitat for species that eventually recolonize a disturbed site (Morrison et al. 2006)
- Influencing patterns of recolonization in disturbed areas (Whelan et al. 2002; Turner et al. 2003)
- Providing a source of energy and nutrients for other organisms (Amaranthus and Perry 1994)
- Modifying or stabilizing environmental conditions in a recovering stand (Perry 1994)
- Shielding young plants from overbrowsing by large herbivorous mammals (Turner et al. 2003)

Each type of disturbance (and, in fact, each disturbance event) leaves its own unique combination of biological legacies (table 2.1). The particular array of legacies present and their composition, pattern, and relative

TABLE 2.1
Major* biological legacies characteristic of different kinds of natural disturbance

Biological Legacies	Type of Disturbance						
	Crown Fire	Surface Fire	Wind-storm	Insect Outbreak	Land-slide	Flooding	Volcanic Deposition
Live canopy trees	+[a]	++	+	++[a]	−	+	−
Understory vegetation & advance regeneration	−	−	+++	++	−	+	−
Standing snags	+++	+	+	+++	−	++	+
Coarse woody debris	++	+	+++	++	++	+	+
Duff, seeds, & rhizomes	+	+	+++	+++	−	+	+
Surviving animals	+[b]	+[b]	+++	+++	−	+[b]	+[b]

*Exclusive of "skips" or areas undamaged or only partially damaged, which can occur in all types of disturbance.
- unusual
+ some
++ many/much
+++ abundant
[a]Depending on species
[b]Surviving in canopy or underground

importance vary with forest type, site, and the behavior and intensity of the disturbance event.

An understanding of the effects of major natural disturbances on the types and densities of biological legacies can help inform ways to better manage forest ecosystems. This information can also help account for historical impacts of past human and natural disturbance. We return to these issues in the following chapters.

Congruence between Natural Disturbance and Human Disturbance Regimes

The biological legacies concept and its relationship to logging impacts (see chapter 3) is closely linked to another pivotal concept in multiple-use forest management—the level of congruence between natural and human disturbance regimes. Many authors contend that if human disturbances (including logging) are similar to natural disturbances, then their effects on ecosystem processes and biodiversity will be more limited (Kuluvainen 2002; Perera et al. 2004). Similarly, Hunter (1993) hypothesized that the maintenance of key ecosystem processes and the conservation of forest biodiversity requires management to be as consistent as possible with natural ecological processes (box 2.3). The logic of these proposals is that, while organisms are best adapted to the disturbance regimes with which they have evolved (Bergeron et al. 1999; Hobson and Schieck 1999; Hunter 2007), they may be susceptible to novel forms of disturbance. Such novel disturbances may be either new combinations of disturbances (Foster et al. 1997; Paine et al. 1998) or disturbances that are more or less frequent and/or more or less intense than would normally occur (Reice 2001) or are completely unprecedented in terms of their selectivity, severity, or scope of action (e.g., exotic herbivores or pathogens).

Disturbance Templates and Variations in Natural Silvicultural Models

Due to the inherent variability in natural disturbance regimes in any given forest ecosystem (and hence the variability in biological legacies that remain after disturbance), no single natural disturbance regime provides a complete model for a silvicultural system. Informed forest

BOX 2.3
The Swedish ASIO Model

If the importance of congruence between natural and human disturbance is accepted, then the scale, pattern, and intensity of logging in managed forests should be broadly similar to the scale, pattern, and intensity of natural disturbance regimes (Bunnell and Kremsater 1990). The Swedes have proposed a model (called ASIO—the acronym refers to the frequency of fire in a particular part of the forest, and it stands for Almost never, Seldom, Infrequent, Often) to use natural disturbance in boreal forests as a template to guide silvicultural practices. The frequency and severity of fire that typically occurs on different terrain provide guidance on rotation length, silvicultural system, and site treatments (fig. 2.9). Similar recommendations have been made for forest ecosystems in neighboring Finland (Niemelä 2003).

- In wetland forest, ravines, and small islands in lakes, there is no forestry activity and fire is excluded.
- Forest on watercourses and flat moist areas historically burned, on average, every 200 years. Uneven-aged management (selection harvesting) and shelterwood harvesting are recommended for these forests.
- Forest in most mesic areas (the majority of boreal forest in Sweden) had burned about every 100 years, the same as the recommended rotation length. It is recommended that cut-over forest be burned and seed trees retained on-site, both practices that had declined over recent decades.
- All dry forestland (e.g., pine forest on sedimentary soils on flat terrain) should be burned approximately every 50 years. These areas support trees that survive recurrent low-intensity fires, so controlled underburning of the forest can be an important tool for maintaining their biological diversity and stand structural complexity, as well as facilitating forest regeneration. Because natural fire regimes resulted in many multi-aged stands, the final felling operation should include the retention of seed trees or uncut patches (Rülcker et al. 1994; Fries et al. 1997; Jõgiste et al. 2005).

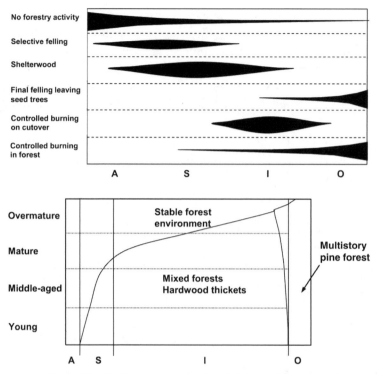

FIGURE 2.9. Variability in fire regimes in Swedish boreal forests and associated variations in harvesting regimes (modified from Rülcker et al. 1994). The acronym ASIO on the x-axis highlights the frequency of fires: A = Almost Never, S = Seldom, I = Infrequent, O = Often. A. The black polygons symbolize the extent of natural disturbance (fire on the x-axis) in relation to human disturbance (on the y-axis).

management will require knowledge of the existence of multiple disturbance pathways and what is, in essence, a continuum of variability (Coates and Burton 1997; Franklin et al. 1997). Therefore, silvicultural prescriptions (including those for salvage logging) will need to vary in different parts of the landscape (Rülcker et al. 1994) as well as within the same stand (Bergeron et al. 1999; Chambers et al. 1999). This is the "don't do the same thing everywhere" approach of Bunnell (1999). Liefers et al. (2003) present a range of silvicultural choices under the categories of age structure, harvesting intensity, site preparation, regeneration materials and methods, vegetation management, and intermediate cuttings. One approach for enhanced ecologically sustainable forest management could be integrating these options with (1) Franklin et al.'s (1997) continuum of variable retention, (2) some version of the variable rotation length model (Curtis 1997; Bergeron et al. 1999; Burton et al. 1999; Seymour and Hunter 1999), and (3) landscape mapping based on both environmental conditions and disturbance regimes. With basic knowledge of disturbance regimes and bounded expectations for the probability of natural disturbances, it then becomes possible to plan for, not just plan around, natural disturbances. These considerations are discussed further in chapters 5 and 6.

The Limits of Natural Disturbance Regimes as Templates and Models

There are limitations to the human–natural disturbance congruence approach. First, it is a hypothesis that not only is difficult to test but actually remains largely untested in most forest ecosystems (Buddle et al. 2006). Second, some very complex processes are extremely difficult to emulate (James and Norton 2002). For example, natural fire patterns are complicated because fire frequencies and fire sizes can be highly variable (Chou et al. 1993; Gill and McCarthy, 1998). Historically, most landscapes were disturbed by indigenous people (King 1963; Bowman 2003; Willis and Birks 2006), making it difficult to determine or agree on the designation of appropriate "natural" patterns (Keith et al. 2002; Spies et al. 2007). Disturbance regimes also change in response to short- and long-term climatic shifts (Emanuel 2005). Bergeron et al. (1998) noted that large changes in the fire frequency in the boreal forests of Canada during the Holocene meant there was no (single) characteristic fire regime for this system. Third, highly targeted actions (that go beyond following natural disturbance regimes) might be needed to meet particular management objectives such as the creation of specialized habitat attributes for an individual threatened species (Simberloff 2004). Finally, few disturbances in modern landscapes are entirely "natural." For example, fires often are extinguished to limit threats to human life and property (Cary et al. 2003).

From a practical perspective, it is clear that human disturbance can never mimic natural disturbance regimes exactly (Haeussler and Kneeshaw 2003; Drever et al. 2006). As an example, wildfires remove only a small percentage of forest biomass because of incomplete combustion of trees (Whelan 1995), in contrast with the large biomass removal in logging (Angelstam 1996). Similarly, while logging removes boles, windstorms and hurricanes simply snap or uproot trees (Foster et al. 1997; Stanturf et al. 2007). The enormous sizes of natural fires that occur, for example, in the boreal forests of North America and the eucalypt forests of southeastern Australia, are considerably larger than would be politically, socially, or ecologically acceptable to use as guides for the sizes of harvest units (Haila et al. 1993).

Given these limitations, an objective of forest research should be to quantify differences between natural and human disturbance regimes and, in turn, to find ways of creating human disturbance regimes more similar (rather than identical) to naturally occurring ones. Indeed, creating an identical replicate of natural disturbance regimes would be impossible (box 2.4) and is not the aim of modified silvicultural systems (Hunter 2007). Despite these problems, natural disturbance regimes are a useful general model for broadly guiding management practices such as logging

BOX 2.4
Contrasts between Logging and Natural Disturbance

Many studies have contrasted the effects on various ecosystem attributes of logging and natural disturbance, especially forest fire (McRae et al. 2001; Haeussler and Bergeron 2004). Such comparisons have been prompted by the often-touted claim that clearcutting and even-aged management are analogous to natural stand-replacing wildfires, and are merely replacing such fires in the life cycle of forests (but see Lindenmayer and Franklin 1997, 2002). Invariably, however, systematic comparisons discover some similarities but many differences between clearcuts and wildfires at many scales (e.g., Lindenmayer et al. 1991; Hobson and Schieck 1999; Buddle et al. 2000; see figure 2.10). For example, there are significant differences in the numbers and spatial patterns of biological legacies that are left in naturally disturbed forests versus clearcut areas. Levels of soil disturbance also differ significantly between naturally disturbed areas and clearcuts (Franklin et al. 2000).

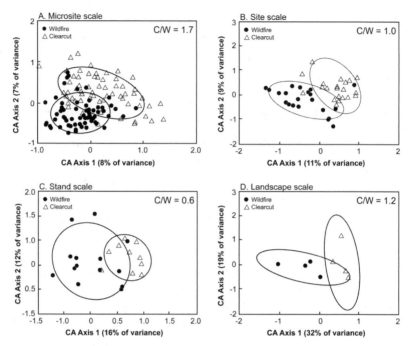

FIGURE 2.10. Multi-scale differences in floristic attributes between mixedwood aspen (*Populus tremuloides*) and white spruce (*Picea glauca*) forests that had been disturbed by wildfire or clearcut logging, as indicated by correspondence analysis (CA) ordination (modified from Haeussler and Bergeron 2004, p. 279). The larger sample ellipses enclose 68% of the range of variability in species composition on the first two ordination axes. The clearcut:wildfire ellipse ratio (C/W) indicates the relative size of the two sample ellipses. Plant species composition 3 years after disturbance indicates greater or equal variability in clearcuts than in wildfires at microsite, site, and landscape scales (C/W ratios ≥ 1.0 in panels a, b, and d, respectively), but less variability than wildfires at the stand scale (C/W ratio < 1.0 in panel c).

(including salvage logging). On this basis, Schulte et al. (2006) and Hunter (2007) recommended that the concept of natural disturbance templates should be integrated with other general principles such as using a range of strategies to ensure a diversity of conditions in stands, landscapes, and regions to best achieve a wide range of management objectives. Haeussler and Kneeshaw (2003) likewise advocated the adoption of a "biocomplexity" approach that maximizes ecological complexity and diversity at all scales.

A General Framework for Characterizing and Comparing Human and Natural Disturbances

A general framework has recently been proposed for contrasting the impacts and implications of both natural and anthropogenic disturbances (Roberts 2004, 2007). Originally devised to explain disturbance effects on understory vegetation, the framework (fig. 2.11) also shows potential for evaluating disturbance effects on wildlife habitat attributes and forest regeneration capability. The framework reflects how the status of the forest canopy, the understory, and soil disturbance can all strongly influence vegetation dynamics and stand regeneration (Brokaw and Lent 1999). Light conditions under a forest canopy vary dramatically if trees are alive or dead, dead but still standing (in which case foliage and fine branches, then successively larger branches and portions of boles are lost), collapsed (in which case woody structure is added to the ground layer), or removed from the site (i.e., harvested). Even after intense crown fires, standing dead trees can ameliorate microclimatic conditions such as sunlight and wind (Perry et al. 1989). Many shade-intolerant tree species are able to regenerate under conditions of 50 percent or more sunlight (Florence 1996).

The input of solar energy to the understory is not a linear function of residual tree cover. Rather, it depends on several factors, including stand maturity, uniformity, stem density, and crown structure. Consequently, local knowledge or calibrations are usually needed to determine the optimal level of canopy removal needed to support understory recruitment by a given tree species. If less than 50 percent of the overstory is disturbed by natural disturbance, however, forest recovery may occur through released growth of surviving trees (Florence 1996) or recruitment of more shade-tolerant tree species (Franklin et al. 2002).

The death or removal of a forest overstory typically creates opportunities for forest understory species, or the release or de novo regeneration of sun-loving tree species. Natural disturbance is so pervasive in the world's forests that most forest ecosystems support several species of relatively fast-growing shrubs, short-lived trees, and herbaceous species that are adapted to take advantage of newly created canopy gaps or stand-level canopy dis-

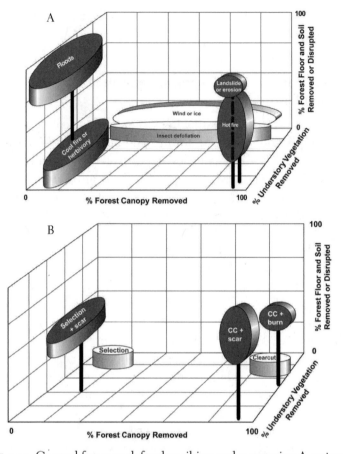

FIGURE 2.11. General framework for describing and comparing A. natural forest disturbances (redrawn from Roberts 2004), and B. first harvest entries of alternative silvicultural systems (redrawn from Roberts 2007). The degree of canopy removal, understory disruption, and soil (or forest floor) disturbance collectively define the disturbance severity in manners informative of changes in ecosystem structure, composition, and function, and facilitates quantitative comparison among different disturbances.

turbance. These understory species typically respond through light-stimulated release after periods of suppressed growth, or by sprouting from rhizomes and buried seeds (e.g., Cumming et al. 2000; Franklin et al. 2002). Consequently, future stand development is greatly affected by whether the understory consists of well-stocked populations of tree seedlings (a "seedling bank") ready to grow into the canopy, or vigorous shrub and herb growth that conversely constrains tree establishment and growth. This undergrowth too may have been affected by the natural disturbance (e.g., in the case of fire), may have been left largely untouched (e.g., in the case of insect outbreaks), or may have experienced patchy disruption, as is often the case when windstorms uproot trees (fig. 2.12).

Despite the overarching importance of light conditions in driving stand dynamics, in some forest types, death or removal of the forest canopy is insufficient stimulus for stand regeneration because thickets of dense understory vegetation or dense mats of organic matter on the forest floor may preclude tree seedling establishment. In many conifer-domi-nated boreal and wet temperate regions, litter decomposition is slow and bryophyte growth on the forest floor is luxuriant. This results in thick or-ganic layers dominating the ground layer (Marcot 1997). When exposed to sunlight (especially full, daylong sunlight), these organic mats (often 10 to 20 centimeters thick) dehydrate (Mallik and Roberts 1994; Oleskog et al. 2000), often making them unsuitable as seedbeds. An important part of describing and comparing forest disturbances, therefore, is the extent to

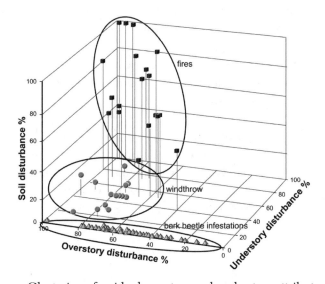

FIGURE 2.12. Clustering of residual overstory and understory attributes in conif-erous forests (west-central British Columbia, Canada) affected by forest fires, wind storms, and bark beetle outbreaks (unpublished data of Ruth Lloyd, Karen Price and Phil Burton).

which the organic layer on the forest floor has been disrupted and the mineral soil exposed (Roberts 2004). The diversity of forest floor substrates is a good predictor of diversity in forest floor plant communities, irrespective of whether it results from smoldering ground fires (Miyanishi 2003), the creation of windthrown root wads (Elliott et al. 2002; James and Norton 2002), or the action of harvesting machinery.

Summary

Natural disturbances are key ecological processes in almost all ecosystems, including forests. In forests, they vary spatially and temporally from frequent, low-intensity, gap-forming disturbances operating at the scale of individual trees to infrequent, landscape-scale, high-intensity events (fires, windstorms, insect attacks) that can significantly alter forest structure. The focus of this book is primarily large-scale, high-intensity disturbances, as these are the kinds of perturbation often followed by salvage logging.

Large-scale, high-intensity natural disturbances can influence many important ecological processes and many elements of biota. They interact with climate and terrain to determine the size, shape, location, types, and content of patches that provide structural complexity and landscape heterogeneity in forest landscapes.

Natural disturbance regimes leave behind a rich array of biological legacies, such as logs, intact thickets of understory vegetation, and large living and dead trees. Biological legacies influence key ecosystem processes (particularly stand-level recovery potential) and provide within-stand structural complexity and habitat for many organisms.

Knowledge and information from natural disturbance regimes at stand and landscape scales can be used to guide the development of human disturbance regimes such as silvicultural systems (including salvage logging practices), so that better outcomes can be achieved for the protection of ecological processes and biodiversity.

Chapter 3

The Effects of Salvage Logging on Ecological Phenomena

Effects of traditional (nonsalvage) logging of green forests have been extensively studied, including such topics as impacts on individual biotic elements, the structure and composition of stands, landscape patterns and composition, and key ecosystem processes and functions (reviewed by Keenan and Kimmins 1993; Hunter 1999; Lindenmayer and Franklin 2002; Burton et al. 2003b). Effects of salvage logging are poorly documented, although there are several important reviews (Beschta et al. 1995, 2004; McIver and Starr 2000, 2001; Kotliar et al. 2002; Karr et al. 2004). These reviews all focus on postfire salvage logging, primarily in the western United States. To the best of our knowledge, equivalent assessments on the effects of salvage logging following windstorms or insect attacks do not exist. A fundamental issue is whether salvage logging, in general, has effects that are different from and potentially more detrimental than logging operations that are not preceded by a major disturbance (Lindenmayer and Noss 2006).

In this chapter, we briefly review the ways in which traditional ("green" tree) commercial logging can affect forest ecosystems and closely linked aquatic ecosystems. We consider why salvage logging might have different, including additional or cumulative, impacts than green tree logging. Our assessment of the published literature on ecological impacts of salvage logging on ecosystem processes, stand structural complexity, landscape heterogeneity, and biota provides the background for a discussion of policy and management implications in chapter 5. We do not consider this review to be comprehensive because much of the research undertaken to date is either unpublished or in "gray literature," which is difficult to access.

The Impacts of Nonsalvage Commercial (Green) Logging

A discussion of the effects of salvage logging must first be preceded by a discussion of nonsalvage commercial logging (hereafter termed green logging). In cases where few if any differences exist between the two, concerns about salvage logging should not extend beyond those concerns that apply to green tree logging impacts.

There is an enormous and highly complex literature on the effects of green tree logging on ecosystem processes and biodiversity. Distinctions are often made between the logging of natural and semi-natural multiple-use forests and the harvesting of plantations. We deal primarily with the former. However, it is well beyond the scope of this book to review in any detail the wide range of impacts of logging natural and semi-natural multiple-use forests.

Very briefly, here are some of the many effects of logging green forests:

- *Altered forest composition.* Harvesting and regeneration aimed at maximizing timber production typically lead to long-term changes in forest composition (Kellas and Hateley 1991; Mueck et al. 1996; Ough 2002). Both tree and nontree components can be altered by harvesting, thinning, or culling (Halpern and Spies 1995). Some species (especially commercial tree species and aggressive understory shrubs and herbs) are often favored, whereas other species decline. Successional trajectories after logging often differ markedly from those that occur after natural disturbances (McRae et al. 2001). In other cases, logging can inadvertently result in the introduction of invasive plant species (Lindenmayer and McCarthy 2001; Brown and Gurevitch 2004).
- *Altered abundance and spatial distribution of structural attributes.* Logging can reduce the structural diversity of stands, such as that associated with live trees, snags, large logs, and thickets of understory vegetation (Lindenmayer et al. 1991; Mladenoff et al. 1993; see box 3.1), as well as alter the spatial distribution of these attributes (Hilmo et al. 2005). Stand structure is often simplified, sometimes drastically, such as when clearcutting is repeatedly utilized (Linden-

mayer and Franklin 1997; fig. 3.1). Animal populations that depend on forest attributes such as cavity trees, fallen logs, and decaying wood are often negatively impacted by commercial logging, especially clearcutting (deMaynadier and Hunter 1995; Lindenmayer et al. 2002; Karraker and Welsh 2006).

• *Altered spatial patterns of vegetation types and stand age classes.* Harvesting in managed forest usually occurs with higher frequency and greater regularity than natural disturbances (McCarthy and Burgman 1995), at least in forest types that are subject to episodic, stand-replacing disturbances. As a consequence of forest regulation and frequent harvesting, natural stand age class distributions are truncated to support uniform areas of younger stands (Seymour and Hunter 1999; fig. 3.3). Those uniform areas also tend to have a smaller size and more discrete edges than those created by natural disturbances (McRae et al. 2001). Changes in the spatial pattern of

---

BOX 3.1
Logging and Declines in Stand Structural Complexity in European Forests

Logging-related declines in the structural complexity of forests have been recognized for a long time, especially in Europe. More than 120 years ago, Gayer (1886) expressed concerns about the simplification of German forests. In Sweden, a century of intensive management in a 123,000-hectare area of boreal forest transformed stand structure from one dominated by widely spaced, large-diameter trees to young, densely stocked forests. The number and volume of large living and dead standing trees in this area was reduced by 90 percent and the extent of old stands by 99 percent (Linder and Östlund 1998; fig. 3.1). It has been recognized that large, dead trees are particularly valuable for many Scandinavian forest taxa (Samuelsson et al. 1994). For example, Berg et al. (1994) calculated that almost 50 percent of the threatened species in Sweden were dependent on dead standing trees or logs. Similarly, mature deciduous trees have also become a rare element of managed stands in Scandinavia (Esseen et al. 1997), although they are a key component of forest composition for a wide range of animal and plant groups (Enoksson et al. 1995).

A

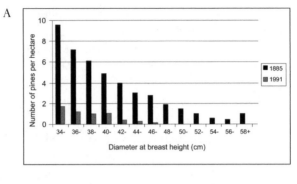

B

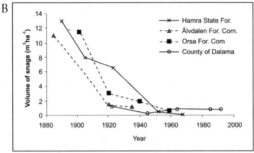

FIGURE 3.1. Altered levels of stand structural attributes in a Swedish forest over the past 200 years (redrawn from Linder and Östlund 1998).

forest age classes may affect populations of some elements of the biota such as large forest owls (Milledge et al. 1991; Lamberson et al. 1994) and some wide-ranging forest mammals (Noss and Cooperrider 1994; Lindenmayer et al. 1999a).

• *Landscape dissection by transportation infrastructure.* Transportation networks constructed for moving logs can provide access to previously remote regions, encouraging the expansion of human populations and associated agricultural practices (Harcourt 1992; Putz et al. 2000) as well as increasing pressure on species that are hunted (Redford 1992; Bull et al. 2001). The World Commission on Forests and Sustainable Development (1999) estimates that between 400 and 2,000 hectares of forest are lost for each kilometer of new road constructed in Brazilian Amazonia.

Roads and railways may also redirect large quantities of runoff and silt into aquatic ecosystems (Naiman and Bilby 1998); increase mortality of animal species through collisions with vehicles (Trom-

A

B

FIGURE 3.2. A. Intensively managed mature forest of Scots pine (*Pinus sylvestris*) in Finland characterized by uniform trees and a sparse understory (photo by Phil Burton). B. Natural old-growth forest of Scots pine in nearby Russian Karelia characterized by multi-layered canopy and abundant undergrowth and lichens (photo by Katya Shorohova).

bulak and Frissell 2000; Forman et al. 2002); facilitate movement of native predators (such as quolls [*Dasyurus vivverinus* and *D. maculates*] in Tasmania) (Taylor et al. 1985), feral predators (e.g., the red fox [*Vulpes vulpes*] and feral cat [*Felis cattus*] in Australia) (May and Norton 1996; May 2001) and weeds or invasive plant species (Wace 1977); and alter habitat connectivity (*sensu* Lindenmayer and Fischer 2006) for particular organisms. For example, culverts under roads change patterns and speeds of water flow in aquatic

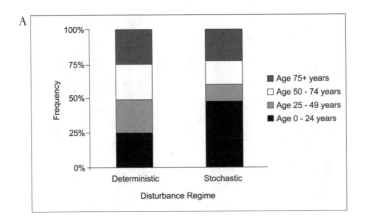

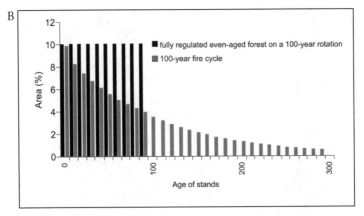

FIGURE 3.3. Differences in the age class distribution of landscape managed for even-aged timber production versus naturally disturbed landscapes. A. Equilibrial states achieved under simulated deterministic (logging) and stochastic (fire) disturbance regimes (redrawn from McCarthy and Burgman 1995); B. Theoretical age class structure of a fully regulated managed forest and a forest landscape naturally disturbed by unselective fires (redrawn from Bergeron et al., 1999).

ecosystems and can provide significant barriers to the movement of some organisms, such as fish (Boubeé et al. 1999; Walker 1999).

Logging and associated forest management practices can directly or indirectly impact key ecosystem processes and forest biota in a myriad of other ways. For example, logging can compact the soil (Rab 1998), and fertilizers applied to promote tree growth or herbicides used to control undesired plants can accumulate in the soil or in hydrological systems (Resource Assessment Commission 1991; White et al. 2006).

Prescribed fire, or slash burning, is a management activity often associated with logging and is used to encourage postharvesting stand regeneration and/or reduce the likelihood and intensity of wildfire (Whelan 1995; but see Moritz et al. 2004). It can influence hydrological regimes, rates of soil erosion, and patterns of nutrient cycling (Van Lear et al. 1985; Carter and Foster 2004; Morley et al. 2004). Prescribed burning also can alter (1) vegetation composition (Haskins and Gehring 2004; Syphard et al., 2006), (2) the spatial pattern of vegetation types, (3) the structure of stands, and (4) the composition of animal communities (James and M'Closkey 2003; Schurbon and Fauth 2003; Noss et al. 2006).

The spatial and temporal patterns of green tree logging can have many influences on forest ecosystems. These influences may be small, for example, where logging involves cutting relatively small areas with the retention of high levels of biological legacies or over very long rotations (Lindenmayer and Franklin 2002). Conversely, short rotations, and high levels of utilization, such as occur through industrial clearcutting and even-aged stand management, can produce numerous homogeneous cutblocks in a landscape. Such outcomes may have significant negative impacts at multiple spatial scales, including landscape-level cumulative effects (e.g., Franklin and Forman 1987).

In summary, nonsalvage commercial (green) logging modifies forest environments in diverse ways over a range of spatial and temporal scales (Hunter 1999; Lindenmayer and Franklin 2002). Some effects of salvage logging are likely to be broadly similar to those we describe for green logging. However, there may be additional impacts.

How Salvage Logging Differs from Green Logging

Salvage logging can differ from conventional logging of green stands in several ways important to the maintenance of ecosystem processes and biodiversity. Four examples are described below.

Conditions Preceding Logging Operations

Major disturbances may be associated with unusual environmental conditions. For example, extensive soil wetting occurs before the high winds associated with hurricanes and cyclones (Elliott et al. 2002). Prolonged droughts and high temperatures are typical before wildfires in some forest types (Wallace 2004), and these can exert strong influences on many organisms (Rübsamen et al. 1984). Consequently, plants and animals are often under stress at the time of disturbance and may not have recovered (or have the potential to recover) from the dual impacts of environmental stress and the disturbance before salvage operations begin.

Conditions Under Which Logging Takes Place

Salvage logging is conducted in disturbed ecosystems where the organic component of soils may have already been burned (Beschta et al. 2004) or the mineral soil exposed (James and Norton 2002). This may make soils more vulnerable to impacts associated with salvage logging, such as compaction and erosion (Shakesby et al. 1996; McIver and Starr 2000, 2001). Salvage logging may also take place around piles of fallen trees, which may make timber removal more difficult (Holtam 1971), thereby requiring more skidding and ultimately increased soil disturbance.

What Is Logged

Salvage logging involves the removal of particular trees or components of stands that are often uncommon in the landscape, such as charred standing stems, insect-killed or dying trees, recently windblown trees, trees partially immersed in volcanic ash, or the largest trees that remain—because of their economic value (Morissette et al. 2002). Conditions following stand-replacing disturbances in many regions are among the most biologically diverse and most uncommon of all forest conditions (Franklin and Agee 2003). Several recent studies show that such ecosystems support distinctive biotic assemblages that are clearly different from those characteristic of other kinds of stands (e.g., Saint-Germain et al. 2004).

Prescriptions for Logging Practices

Salvage logging sometimes occurs in ways that are more intensive at the stand level or extensive at the landscape level than green logging (McIver and Starr 2000; Schmiegelow et al. 2006; Victoria Department of Sustainability and Environment 2007). Salvage logging may also be allowed in areas where green logging would normally be prohibited (Foster and Orwig 2006). The negative consequences are many:

- Cutblock sizes can be larger (British Columbia Ministry of Forests and Range 2006).
- Forests may be cut at much younger ages than normal (Radeloff et al. 2000).
- Larger or older trees may be removed when it is not otherwise allowed (Thrower 2005).
- Larger quantities of slash may be left behind and sometimes burned due to greater tree breakage, defects, or smaller tree size (fig. 3.4).
- Areas previously designated as roadless may be roaded, providing access for potential future logging (Karr et al. 2004).
- Road networks may be more extensive and more intensively utilized (British Columbia Ministry of Forests and Range 2006); but they also may not be constructed to high engineering standards.
- Particular kinds of trees, stands, or areas normally reserved from logging (e.g., old-growth reserves) may be logged (Eggler 1948; Bunnell et al. 2004; Foster and Orwig 2006).

FIGURE 3.4. Piles of waste wood (small trees, tops, and logs culled because of cracks) remaining from salvage of beetle-killed lodgepole pine (*Pinus contorta*) in central British Columbia (photo by Phil Burton).

- Areas may be logged at times when it is otherwise not allowed. For example, restrictions on logging during the wet season were lifted as part of salvage after the 2002 Biscuit Burn in southern Oregon (DellaSala et al. 2006b).

As an example of altered prescriptions, in the tropical rain forest of Indonesia, fire-damaged forests can be re-entered for salvage logging at shorter intervals than unburned stands (van Nieuwstadt et al. 2001). As a result of extensive and intensive salvage logging efforts following widespread infestations by the mountain pine beetle in interior British Columbia (British Columbia Ministry of Forests and Range 2006), the overall rate of forest harvesting on public lands has more than doubled in some districts, and the density of logging roads and the level of road use have increased significantly (fig. 3.5).

The Ecological Impacts of Salvage Logging

Because salvage logging is a combination of disturbances, there is a fundamental question whether salvage logging has different and/or additional effects than either a natural disturbance alone or green logging alone. The answer to this question is elusive due to the limited study of salvage logging impacts (Morissette et al. 2002) and the limitations inherent to many existing studies (McIver and Starr 2000), including differences in the scale (patch versus stand versus landscape) at which effects are measured.

Information accumulated to date suggests that the impacts of salvage logging will vary in response to a wide range of factors, just as in the case of green logging:

- Type, intensity, frequency, and spatial pattern of logging and the preceding disturbance (e.g., windstorm versus wildfire versus insect attack). The spatial and temporal magnitude of large-scale, high-intensity salvage operations is likely to be different from those of green logging operations.
- Levels and types of biological legacies that are retained (Macdonald 2007)
- Specific nature of the ecosystem, ecological processes, and biota involved. Impacts of salvage logging, for example, are likely to be dif-

A

B

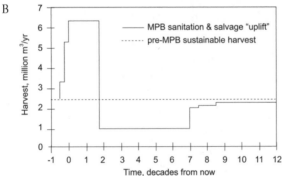

FIGURE 3.5. A. Salvage logging under way near Prince George in interior British Columbia (photo by Ken Hodges). The overall rate of harvest is much higher than permitted for non-salvage commercial harvesting, as shown in B. Timber supply projection showing uplift in rate of harvest to salvage beetle-killed timber, followed by a fall-down in timber availability (data for Quesnel timber supply area courtesy of B.C. Ministry of Forests and Range).

ferent between burned old-growth stands and burned young re-growth (second-growth) stands.
• Combined impacts of the type and intensity of a preceding disturbance and a specific intensity and pattern of logging (Odion et al. 2004; Thompson et al. 2007)

- Post-disturbance weather conditions and their influence on mass movements of debris (Helvey et al. 1985) and soil erosion generally (McIver and McNeil 2006)

Ecological impacts of salvage logging can be classified into two broad categories based on existing research: impacts on organisms and habitats, including impacts on the physical structure of forest stands and aquatic systems; and impacts on key ecosystem processes (e.g., hydrological cycles, nutrient cycling, and soil formation). These impacts are considered in more detail below. They are often interrelated, such as when ecological processes influence the persistence of individual species (James and Norton 2002). Impacts also can be additive or cumulative; the potential for such kinds of effects is discussed in a subsequent section.

Salvage Logging Impacts on Organisms and Habitats

Patterns of ecosystem recovery together with the recovery of many biotic elements of the biota within forest ecosystems are strongly influenced by the types, numbers, and spatial arrangements of biological legacies that remain following disturbances (Lindenmayer and Franklin 2002; see chapter 2; table 3.1). Biological legacies are critical for biodiversity and influence the rate and pathway of post-disturbance recovery (Platt and Connell 2003; Mazurek and Zielinski 2004; Hyvärinen et al. 2006). Salvage logging explicitly removes some of the biological legacies (i.e., tree boles) generated by disturbances, thereby potentially diminishing some or all of the ecological roles they play (which are listed in chapter 2). Removal of biological legacies also simplifies the structure of ensuing forest stands (Franklin et al. 2000; Titus and Householder 2007), homogenizes landscape pattern (Radeloff et al. 2000; Crisafulli et al. 2005), and reduces habitat and landscape connectivity between undisturbed areas (Morissette et al. 2002).

As an example, wildfires in the montane ash forests of the Central Highlands of Victoria (southeastern Australia) led to about 70 percent of stands being even-aged regrowth forests. The remaining 30 percent of stands are multi-aged and comprised of trees from multiple cohorts (McCarthy and Lindenmayer 1998); they occur primarily on south-facing slopes with low levels of incoming radiation and on flat terrain (Mackey et al. 2002). However, salvage logging has reduced the prevalence of multi-aged stands from

TABLE 3.1

Examples of salvaging logging impacts or potential impacts on organisms, stand structure, and landscape composition

Impact	Location	Citation
Reduced nesting and foraging habitat for vertebrates	Quebec, Canada	Nappi et al. (2003)
	Alberta, Canada	Hobson and Schieck (1999)
	Montana and Wyoming, USA	Hutto (1995, 2006), Hutto and Gallo (2006)
	Washington State and Oregon, USA	Bull et al. (2001), Wales (2001)
	Alaska, USA	Murphy and Lehnhausen (1998);
	Idaho, USA	Saab and Dudley (1998), Russell et al. (2006), Saab et al. (2007)
	Victoria, Australia	Van der Rhee and Loyn (2002), Lindenmayer and Ough (2006)
	Israel	Haim and Izhaki (1994)
Altered bird community	Saskatchewan, Canada	Morrisette et al. (2002)
	Montana and Wyoming, USA	Hutto (1995; 2006), Hutto and Gallo (2006)
	Alaska, USA	Murphy and Lehnhausen (1998)
Altered in-stream macroinvertebrate community	Western USA	Karr et al. (2004)
Reduced vegetative recovery and altered plant species composition	Oregon, USA	Isaac and Meagher (1938), Donato et al. (2006a, 2006b)
	Washington State, USA	Titus and Householder (2007)
	California, USA	Stuart et al. (1993)
	Eastern USA	Frothingham (1924)
	Northeastern USA	Cooper-Ellis et al. (1999), Kizlinski et al. (2002)
	Victoria, Australia	Lindenmayer and Ough (2006)
	Kalimantan, Indonesia	van Niewstadt et al. (2001)
	Quebec, Canada	Purdon et al. (2004); Greene et al. (2006)
	Korea	Che and Woen (1997)
Altered patterns of landscape heterogeneity	Wisconsin, USA	Radeloff et al. (2000)
	Washington State, USA	Crisafulli et al. (2005)
	Northeastern USA	Orwig et al. (2002)
	Victoria, Australia	McCarthy and Lindenmayer (1998)

30 percent to 7 percent and altered the spatial pattern of these kinds of forests because flat areas are readily accessible. These changes in stand structure and landscape pattern have had corresponding negative impacts on biodiversity, including rare species such as Leadbeater's Possum (*Gymnobelideus leadbeateri*; Lindenmayer and Ough 2006; see chapter 4).

Dead and charred trees created by wildfires and other natural disturbances are biological legacies that can be eliminated or depleted by salvage logging (Murphy and Lehnhausen 1998; Nappi et al. 2003; fig. 3.6). For example, Russell et al. (2006) found that salvage logging in burned forests in the U.S. state of Idaho reduced both snag abundance and overall snag diameter compared with burned areas that were not salvage logged. Another study found that snag biomass after salvage logging was

FIGURE 3.6. Large burned trees with hollows survive for decades in Victorian montane ash forests. These trees are key biological legacies for more than 40 species of cavity-dependent vertebrates (photo by David Lindenmayer).

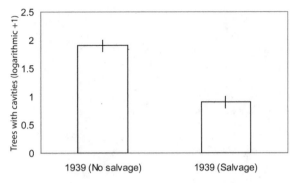

FIGURE 3.7. Depletion in the abundance of large cavity trees (per 3-ha site) in salvage logged montane ash forests of the Central Highlands of Victoria. Data are from field surveys of 376 3-ha sites throughout the Central Highlands of Victoria. Cavity trees are key biological legacies for arboreal marsupials and other hollow-dependent vertebrates in these forests (see Chapter 4).

FIGURE 3.8. Leadbeater's Possum (*Gymnobelideus leadbeateri*) is a nationally endangered Australian species for which the prolonged depletion of large fire-damaged trees by salvage logging has a two-hundred-year negative impact (photo by David Lindenmayer).

approximately half that in burned but unsalvaged forest (McIver and Ottmar 2007). The depletion of biological legacies has major implications for biota dependent on them (Hutto 1995, 2006; Saab and Dudley 1998; Haggard and Gaines 2001; figs. 3.8 and 3.9).

Other kinds of structural features that can be depleted or eliminated by salvage logging include malformed trees (Cooper-Ellis et al. 1999),

FIGURE 3.9. Burned trees in fire-damaged habitats are key habitats for some species of woodpeckers such as the black-backed woodpecker (*Picoides articus*, photo by Richard Hutto).

trees with large epicormic branches (Hanson and North 2006), large logs and coarse woody debris (Minshall 2003), and tip-up mounds (Cooper-Ellis et al. 1999). Levels of vegetative reproduction (suckering and stump sprouting) and leaf area recovery also may be significantly affected by salvage logging (Fraser et al. 2004).

Major natural disturbances provide important pulses of biological legacies such as large-diameter dead standing trees and large pieces of coarse woody debris (Horton and Mannan 1988; Stanturf et al. 2007), including the recruitment of coarse woody debris into streams and other aquatic environments where they provide habitat and nutrients for a wide range of organisms (Minshall 2003). As in the case of green logging, unless significant retention of large woody structures is planned, salvage logging will be followed by prolonged periods of time before new ones are created (Chambers and Mast 2005). Removal of burned standing trees obviously precludes their recruitment to the forest floor and streams (Minshall 2003). Lindenmayer and Ough (2006) speculated that the loss of large-cavity trees in salvage logged montane ash forests in Victoria will take at least two hundred years to reverse. This is because cavity formation in Australian eucalypts is an extremely slow process in the absence of primary excavators such as woodpeckers. It is facilitated primarily by the activities of fungi and termites (Gibbons and Lindenmayer 2002).

Impacts of salvage logging on organisms and stand structure have been examined in several studies. Results are variable, as would be expected from work conducted across a range of forest types subject to different intensities and frequencies of human and natural disturbance (e.g., Stuart et al. 1993; Greene et al. 2006; McIver and Ottmar 2007) and on different taxa among and within forest types (e.g., Greenberg et al. 1994; Greenberg and McGrane 1996). Research on altered stand structural features and biota associated with dead and charred trees (particularly birds and mammals) is most common, with many of these studies demonstrating or forecasting negative impacts from salvage logging (e.g., Morissette et al. 2002; McIver and Ottmar 2007). Neutral or positive outcomes have been reported in some investigations (Greenberg et al. 1995a, 1995b; Greenberg 2002; Macdonald 2007), such as those of microbial assemblages (Khetmalas et al. 2002) and plants (Ne'eman et al. 1997; Elliott et al. 2002). Area-sensitive species, such as the sharp-tailed grouse (*Tympanuchus phasianellus*), in the Pine Barrens of northwestern Wisconsin (USA) benefit from the extensive open habitats created by salvage logging (Radeloff et al. 2000). Similarly, foraging conditions for large mammals such as the grizzly bear (*Ursus arctos horribilis*) may be improved by salvage logging, although additional roads constructed to access timber might expose the species to increased human hunting pressure (Bunnell et al. 2004).

Salvage logging may have other direct impacts on biodiversity. Post-disturbance plant recovery can be changed (e.g., levels of resprouting and germination from seeds; Cooper-Ellis et al. 1999; Lindenmayer and Ough 2006), leading to low levels of plant recruitment (Greene et al. 2006) and altered plant species composition and abundance of plant species and life-forms (Stuart et al. 1993; Greene et al. 2006). For example, seedlings that germinate following a wildfire may be damaged or killed by mechanical disturbance associated with subsequent salvage logging (van Nieuwstadt et al. 2001; Fraser et al. 2004). Natural regeneration of conifers after high-severity fire in southwestern Oregon (USA) was generally abundant in unsalvaged stands, but reduced in salvage logged areas

(Donato et al. 2006a, 2006b). Salvage logging has significantly reduced recruitment of two conifer species in the Canadian province of Quebec, whereas post-salvage densities of trembling aspen (*Populus tremuloides*), which reproduces asexually, were not affected (Greene et al. 2006). Salvaging burned trees reduced seedling recruitment and rates of tree growth in conifer stands in southeastern Spain, although impacts were not deemed to be substantial where initial levels of seedling establishment were high (Martínez-Sánchez et al. 1999).

Bycatch and Conversion Impacts

Bycatch is another significant effect that can result from salvage logging. It not only constitutes another example of the removal of biological legacies but can dramatically alter plant species composition. Bycatch is a term borrowed from fisheries management, but here refers to the removal of trees and other vegetation that is not directly impacted by a disturbance. It is a significant issue in interior British Columbia, where mixed stands of damaged and undamaged tree species are being harvested as part of extensive salvage logging operations over broad areas of lodgepole pine killed by the mountain pine beetle (Bunnell et al. 2004; Burton 2006). Similar issues arise in the northeastern United States, where hemlock woolly adelgid (*Adelges tsugae*) is killing stands of eastern hemlock (*Tsuga canadensis*)—a tree species of low economic value—and other tree species are being cut to fund salvage logging operations (Brooks 2004; Foster and Orwig 2006). We discuss these examples in further detail in chapter 4.

Salvage logging can lead to large-scale land-use conversion. In the southeastern United States, many areas of native forest were converted to softwood plantations following salvage logging operations after Hurricane Hugo (Conner et al. 2004).

Salvage Logging and the Degradation of Early Successional Habitats

Naturally regenerating postfire communities that support structural biological legacies are uncommon in some landscapes (Franklin and Agee 2003) as a consequence of long-term fire suppression, historical logging practices, and aggressive salvage logging policies (Zackrisson 1977; Shinneman and Baker 1997; Society for Conservation Biology Scientific Panel on Fire in Western U.S. Forests 2005). Salvage logging removes key structural and other attributes from early successional habitats and thereby negatively affects species closely associated with them, substantially altering assemblages and communities (Morissette et al. 2002). This can result in serious problems for species associated with such habitats (Purdon et al. 2004). This is the case for many fire-associated endangered species in Scandinavian forests (Heliövaara and Väisänen 1984; Wikars 2001) where salvage logging of burned areas has removed much of the early successional habitat needed by them. Early successional natural habitats are notably different from young logged and regenerated stands in species composition, structural features, and a wide range of other attributes. For example, numerous studies have highlighted differences between post-disturbance conditions following wildfire and those following clearcutting (Lindenmayer et al. 1991; McRae et al. 2001; Stuart-Smith et al. 2006).

Natural recovery of ecosystems following disturbance is important for many ecological processes as well as individual species. Fail (1999) suggested that rates of recovery of forest in the U.S. state of South Carolina salvaged after Hurricane Hugo might have been faster if salvage logging had not occurred. In the case of fire-damaged stands and subsequently salvage logged stands, human intervention to speed the re-establishment of closed-canopy forests has been highly controversial (Beschta et al. 2004; Shatford et al. 2007). Deliberate planting to restore tree cover may consequently do the following:

- Impair natural regeneration of local species and eliminate local genotypes (Ne'eman et al. 1997), including those (harvested as by-catch) which might otherwise have been disturbance-resistant.
- Reduce or eliminate the distinctive biodiversity and stand structure of slowly recovering forest.
- Alter natural patchiness and heterogeneity in vegetation cover, by creating, for example, dense homogeneous stands (Crisafulli et al. 2005).
- Create dense stands where they are ecologically inappropriate, increasing the risk of high-intensity wildfires and subsequent reburning (Odion et al. 2004; Thompson et al. 2007).
- Result in the introduction of invasive plants such as exotic grasses and other herbaceous species (Beschta et al. 2004). These may, in

turn, alter fuel levels and promote future fires (Robichaud et al. 2003; Odion et al. 2004) or slow the recovery of native plants (Turner et al. 2003).

Issues related to human intervention to speed post-disturbance "recovery" or forest re-establishment characterize debates about salvage logging and planting in the Douglas-fir (*Pseudotsuga menziesii*) forests of the U.S. Pacific Northwest (Franklin 2004, 2005, 2006; Shatford et al. 2007). Large, naturally regenerating areas that have not been subject to salvage and subsequent replanting (e.g., figure 3.10) are uncommon (Franklin and Agee 2003). An examination of a 5.6-million-hectare forest land base in central British Columbia reveals that only 61,000 hectares (1 percent) originated from unsalvaged fires that occurred less than sixty years ago (P. Burton, unpublished data). Although partly a record-keeping issue, this is much less than would be expected in unmanaged landscapes. These habitats are valuable, indeed essential, for many species (Society for Conservation Biology Scientific Panel on Fire in Western U.S. Forests

FIGURE 3.10. Slow-recovering post-fire habitat is uncommon in many parts of the Pacific Northwest of the USA. This high-elevation burn in the Okanogan National Forest is still dominated by a diversity of plant life forms and avifauna over a century after a fire (photo by Jerry Franklin).

2005). Early successional, structurally complex environments recovering following fire have been shown to be important in Canadian forests, and they support distinctive assemblages of organisms ranging from birds (Nappi et al. 2004) to beetles (Saint-Germain et al. 2004).

Salvage Logging Impacts on Ecosystem Processes

In unmanaged ecosystems and landscapes, major natural disturbances generate heterogeneity by partially resetting the successional clock and reducing the potential for climax species to prevail (Collins et al. 1995). Interestingly, small- or gap-scale disturbances also generate heterogeneity within dense young tree stands that develop homogeneous internal structure through competitive processes (Franklin et al. 2002). All disturbances can enhance some ecological processes such as primary productivity and nutrient cycling through the pulsed release of resources (Bazzaz 1983).

Major disturbances also can aid ecosystem restoration by creating some of the structural complexity and landscape heterogeneity lost through past human management. For example, floods can reshape riparian areas through movement of sediment and other debris (Bayley 1995). Floods can also restore levels of primary productivity (Johnson et al. 1994) such that they revitalize human-modified aquatic ecosystems (Gregory 1997). Hurricane Katrina is believed to be responsible for rejuvenating the sandy bottoms of streams of the coastal plain of parts of the southern United States (Adams 2006). Similarly, wildfires generate dead wood and promote the development of cavities in trees (Inions et al. 1989)—structural attributes severely depleted by some forestry practices.

Wildfires can create structural complexity and landscape heterogeneity lost through past management. They may also reverse the process of paludification, or development of deep organic layers and perched *Sphagnum* bogs, and release nutrients otherwise unavailable (Fenton et al. 2006). Another example of rejuvenation is the effect of windthrow in retarding development of iron pans, perched water tables, and paludification in coastal forests in southeastern Alaska (Bormann et al. 1995) and adjacent British Columbia (Banner et al. 2005).

Salvage logging has the potential to alter many important ecosystem processes (table 3.2). Hydrological regimes are an example, and they can be modified via altered overland flow rates, soil erosion, and consequent

TABLE 3.2
Examples of salvage logging impacts or potential impacts on ecosystem processes

Impact	Location	Citation
Altered hydrological regimes	New England, USA	Foster et al. (1997), Foster and Orwig (2006), Reeves et al. (2006)
	Northwestern USA	Beschta et al. (2004), Karr et al. (2004)
	South Carolina, USA	Amatya et al. (2006)
Increased sediment flows in watersheds and coastal zone environments	Western USA	Karr et al. (2004)
	Oregon, USA	McIver and McNeil (2006)
	Portugal and Spain	Shakesby et al. (1993, 1996)
	Southeast Asia	Lindenmayer and Tambiah (2005)
Reduced soil nutrient levels	Quebec, Canada	Brais et al. (2000)
	California, USA	Johnson et al. (2005)
	Northeastern USA	Kizlinski et al. (2002)
	Washington State, USA	Titus and Householder (2007)
Disturbed soil layers and increased soil compaction	Maine, USA	Hansen (1983), M. Hunter (personal communication)
Reduced shading, increased soil temperature, and soil compaction	Quebec, Canada	Purdon et al. (2004)
Increased edge effects	California, USA	Hanson and Stuart (2005)
Increased fine fuels contributing to short-term fire risk	Oregon, USA	Donato et al. (2006a).
Increased stem density resulting from post-salvage planting which in turn elevates risks of reburning and future fire intensity	Oregon, USA	Thompson et al. (2007)

in-stream sedimentation (Mackey and Cornish 1982; Helvey 1980; Helvey et al. 1985; McIver and McNeil 2006). Salvage logging activities also limit cavity-tree formation and availability (Lindenmayer and Ough 2006; Russell et al. 2006), soil profile development (Greene et al. 2006), and nutrient cycling (Johnson et al. 2005). As noted by Cooper-Ellis et al. (1999, 2693), in contrast to the natural recovery of a disturbed ecosystem, salvage logging has the potential to "convert a relatively intact system to a strongly modified site in which ecosystem control is reduced."

An example of the potential for salvage logging to impair ecosystem processes is the prolonged change in regional hydrological regimes that occurred after the massive "cleanup" that followed the unnamed 1938 hurricane in the U.S. Northeast (Foster et al. 1997; see figure 4.15 and one of the case studies in chapter 4).

Results of a long-term study of paired watersheds in the U.S. state of South Carolina (Amatya et al. 2006) revealed that streamflows (expressed as a percentage of precipitation received) were higher in the watershed where salvage logging had taken place after Hurricane Hugo in 1989 than in the nearby watershed where there was no salvage logging. In Portugal, postfire salvage logging and subsequent site preparation for replanting led to sediment losses 100 times greater than background levels (Shakesby et al. 1993). The sediment-catching role played by logs, leaf litter, and understory vegetation is lost when they are removed. This can result in significant negative impacts of subsequent salvage logging operations on aquatic ecosystems and associated macroinvertebrates (Minshall 2003). Such effects are apparent in burned watersheds in southeastern Australia where extensive salvage logging of exotic radiata pine (*Pinus radiata*) plantations has occurred (Lindenmayer 2006; White et al. 2006; see chapter 4). McIver and Starr (2000, 2001) concluded that additional road building associated with salvage logging and ground-skidding of logs (which altered the properties of upper soil layers) increased both soil compaction and erosion in already fire-damaged watersheds. The horizon depth and organic content of soils were significantly more affected by subsequent fires on salvaged areas of windblown forests in the U.S. state of Maine than on unsalvaged areas (Hansen 1983; M. Hunter, personal communication). Salvage logging may severely damage water-repellent soils that develop following burning (Beschta et al. 2004). This, in turn, may accelerate levels of soil erosion with corresponding impacts on aquatic ecosystems. In general, preliminary research suggests that intact stands of dead trees have hydrological properties (related to the interception of precipitation, shading, and intact understories) intermediate between those of mature living forests and those subject to clearcut logging (Redding et al. 2007). This watershed protection role can be particularly important in some catchments, depending on overall basin structure, cover types, and disturbance levels.

Salvage logging affects levels of soil nutrients and, consequently, patterns of nutrient cycling. Studies in Quebec, Canada, show that salvage logging on sites subject to high-severity fires led to depletion of soil calcium, magnesium, and phosphorus such that levels of these nutrients would not return to prefire levels within the planned rotation time of 110 years (Brais et al. 2000). Similar work in the U.S. state of Nevada revealed that soils in salvaged ecosystems had significantly less carbon and more nitrogen than soils under adjacent unsalvaged areas (Johnson et al. 2005). Conversely, Beschta et al. (2004) speculated that postfire salvage logging that damages nitrogen-fixing plants may reduce rates of nutrient replenishment. Fail (1999) reasoned that in forest stands subject to major windthrow events, decomposing biomass in unsalvaged areas coupled with increased light levels resulting from altered canopy conditions would increase rates of net primary productivity and, in turn, more rapid rates of forest recovery. However, he did not find statistically significant support for this hypothesis. Work on the effects of the 1950 blowdown event on watersheds in the Adirondack Mountains of New York State (USA) suggested that extensive windthrow contributes to significant acidification of some lakes (Dobson et al. 1990). Conversely, the authors concluded, "Salvage logging appears to counteract the effects of blowdown and facilitate short-term [lake] acidification followed by long-term neutralization" (Dobson et al. 1990, 357).

The removal of biological legacies such as large standing and fallen trees has the potential to alter the spread, pattern, and intensity of subsequent fires in salvage logged stands. The impacts of salvage logging on fuels and fire regimes are important because fire is a key ecological process in many ecosystems (chapter 2). Some workers have argued that dead trees and fallen logs left following natural disturbance create abundant fuels and an increased fire risk (Sessions et al. 2004; Passovoy and Fule 2006). This a major concern in large parts of interior British Columbia where extensive stands of dead lodgepole pine have resulted from infestations of mountain pine beetle (Hughes and Drever 2001; British Columbia Ministry of Forests and Range 2006). However, salvage logging may increase rather than reduce fire risk (Thompson et al. 2007). This was one of the conclusions of a much-discussed study by Donato et al. (2006a) in southern Oregon (USA). Their data (albeit limited) suggest that salvage logging can significantly increase fine and coarse fuels, leading to a short-term elevation in fire risk (fig. 3.11). Odion et al. (2004) and Thompson et

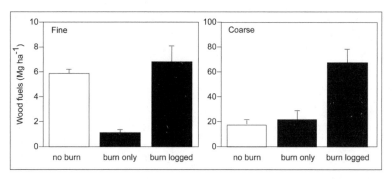

FIGURE 3.11. Contrasts in fine (≤7.62 cm diameter) and coarse (>7.62 cm diameter) fuels subject to different treatments (redrawn from Donato et al. 2006a).

al. (2007) reached similar conclusions about altered fuels and increased severity of subsequent fires. High stand density created by planting trees closely together after salvage logging was considered to be the mechanism contributing to increased risks of reburning (Thompson et al. 2007). Notably, large pieces of coarse woody debris that are recumbent on the forest floor typically contain high levels of moisture (Harmon et al. 1986) and may support well-developed mats of bryophytes (Ashton 1986; Lindenmayer et al. 1999b). Under these conditions, large pieces of highly decayed coarse woody debris can act as microfirebreaks in forests (Campbell and Tanton 1981; Andrew et al. 2000). Conversely, a forest stand disturbed several years previously may be characterized by a maze of "jack-strawed," elevated, air-dried logs that can burn very intensely if a fire spreads from elsewhere (Kulakowski and Veblen 2007), although such accumulations of large fuels are not likely to ignite readily.

Salvage logging can have other kinds of unexpected impacts on ecosystem processes. A recent study in northwestern California showed that salvage logging significantly altered edge environments in comparison to edges created by wildfires but which remained unsalvaged (Hanson and Stuart 2005). In particular, the depth of edge influence was increased by 15–30 meters in salvaged stands. Plant species composition, in particular the abundance of old-growth–related plant species, was altered at salvage edges and contrasted markedly with natural (nonsalvage) fire-created edge environments (Hanson and Stuart 2005). More generally, the removal of trees and disruption to the understory amplify the contrast between disturbed and undisturbed forest. Clearcut salvage logging creates sharp boundaries, thus increasing edge effect impacts beyond those typically associated with natural disturbances (Harper et al. 2005).

Salvage Logging and Carbon Budgets

With global climate change now generally accepted (Millennium Ecosystem Assessment 2005; Pittock 2005), there is a growing awareness of the impacts of human actions on the accelerated release of greenhouse gases such as carbon dioxide and methane into the atmosphere. Forest management can sequester carbon and thereby prevent or offset some of the worldwide carbon dioxide pollution. In this context, how does salvage logging after disturbance affect the net uptake and fixation of carbon?

Forest disturbances causing tree damage and death result in pulses of biomass (mostly cellulose, which is approximately 50 percent carbon) available for decomposition in an ecosystem. Depending on the nature and severity of disturbance, and the physical size of individual components, biomass may be destined for rapid or slow decomposition (Kurz and Apps 1999). For example, it may take months or years for trees to die after an insect attack or windstorm, for leaves and twigs to fall to the ground, and for this material to gradually be consumed (decomposed) by bacteria and fungi that then release carbon dioxide through respiration. Even after intense, stand-replacing crown fires, approximately 60 percent or more of the coarse woody debris and almost all standing tree boles greater than 3 centimeters in diameter remains unburned (Stocks et al. 2004) and begins to slowly decay. The speed at which most biomass decomposes depends on moisture, temperature, and ease of colonization by several guilds of decomposers, such that logs lying on the ground typically break down more rapidly than elevated logs or standing dead trees (Rice et al. 1997; Naesset 1999; Mackensen et al. 2003). Different agents and severities of disturbances, and different salvage practices, further result in different proportions of woody biomass on the ground. Krankina and Harmon (1995) conclude that the general regime of intensive forest management (including short rotations, thinning, and salvage operations) in boreal Europe reduces dead wood carbon stores to 5–40 percent of levels found in undisturbed old-growth forest, while natural disturbances increase the dead wood carbon pool by a factor of two to four.

The degree to which salvage logging alters the natural release of carbon depends on the severity of disturbance and the intensity of salvage logging (and how they each partition biomass into slow- and fast-decomposing pools). It further depends on how the site-specific potential for natural recovery and regrowth is altered. As discussed in chapter 2, most natural disturbance events are highly variable, leaving living trees and/or understory vegetation behind in varying degrees and patterns. Those surviving plants, and those that invade and establish after the disturbance, may rapidly match the growth rates found in the predisturbance forest. At the scale of 1 square kilometer pixels for fires in the boreal forest of Canada, it is estimated that it takes about eighteen to twenty years for forests to recover to predisturbance levels of productivity (Amiro et al. 2000). In contrast, postfire productivity in Australian forests dominated by *Eucalyptus* spp. capable of epicormic sprouting recovers in as little as two or three years (Chattaway 1958). Hence, it is also important to consider the impacts of salvage logging on the distinctive recovery trajectory following disturbances in different forests. In addition, carbon balance calculations must include how much fossil fuel is consumed in the harvest, transport, and processing of wood, and how much of the resulting wood products (and waste products from harvesting and manufacturing) are channeled into long-term (e.g., structural timber) or short-term (e.g., newspaper and toilet paper) carbon pools.

Few studies have investigated the carbon balance implications of salvage logging, given the complexity of such estimates. Fail (1999) found that average levels of soil organic matter were 50 percent greater in unsalvaged stands than in salvaged stands two to six years after Hurricane Hugo in the U.S. state of South Carolina. However, there were no significant differences in area-based productivity or decomposition rates measured over the same period. McNulty (2002) evaluated the overall impacts of hurricanes and posthurricane salvage logging across the eastern United States. Despite increased soil nutrient levels and growth rates in surviving trees, he concluded that reduced stocking and delays in leaf area recovery result in area-based forest productivity taking fourteen to nineteen years to recover to predisturbance levels after a typical hurricane. He noted that 13 to 15 percent of damaged or killed trees may be salvaged, but since stemwood averages 64 percent of the total carbon content of a tree, only 9 percent of the total carbon pulse is channeled into forest products, the rest being allowed to decompose in the forest (McNulty 2002). McNulty (2002) pointed out that the fate of unsalvaged debris and logging slash can also influence the carbon balance—much carbon can rapidly be lost to the atmosphere if this material burns. It remains to be demonstrated how the net carbon balance of the entire process of logging, transport, reforestation, waste disposal, and forest product manufacturing compares with that associated with allowing hurricane-damaged forests to recover naturally.

The Carbon Budget Model for the Canadian Forest Sector (Canadian Forest Service 2007) incorporates user-specified partitioning of biomass to different organic matter pools after natural disturbance and logging, and includes flexible stand-recovery trajectories (Kurz et al. 2002). Preliminary simulations of the mountain pine beetle outbreak in British Columbia indicate considerable short-term carbon losses associated with an aggressive salvage logging policy (W. A. Kurz, personal communication). A systematic comparison of the carbon budget implications of different disturbances and associated salvage logging in forests around the world has not yet been done. But it is clear that accelerated rates of forest disturbance, whether by natural agents or logging or both, result in periods of time over which forests can be a net source rather than a net sink of carbon (Kurz and Apps 1999; Law et al. 2004).

Intersection of Impacts on Organisms and Ecological Processes

We have presented information on impacts of salvage logging on organisms and ecological processes in separate sections. Of course, such impacts are almost always intimately intertwined. Salvage logging of riparian vegetation can alter ecohydrological processes such as in-stream microclimatic regimes and rates of sedimentation (Karr et al. 2004; Reeves et al. 2006). This, in turn, can have significant additional impacts on aquatic biota (Bunnell et al. 2004). Similarly, modification by salvage logging of ecological processes such as fire regimes, cavity development, and soil formation will have impacts on a wide range of taxa strongly influenced by these processes (Lindenmayer and Noss 2006). Haim and Izhaki (1994) speculated that salvage logging may change microclimatic conditions and this, in turn, might be the reason for the contrasting rodent assemblages they observed in the salvaged and unsalvaged Mediterranean pine (*Pinus halepensis*)–dominated ecosystem they studied in Israel. Perry et al. (1989) suggested that removal of burned trees will result in the loss of shading effects that are important for the regeneration of plants on dry and higher elevation sites. Similarly, Purdon et al. (2004) speculated that soil drying due to burned snag removal was one of the ecological processes that underpinned altered understory plant species abundance, composition, and richness associated with salvage logging in a Canadian boreal forest. They also considered the process of soil compaction to have had a major impact on plant species composition (Purdon et al. 2004).

Salvage Logging and Potential Cumulative Effects

Organisms are typically best adapted to the disturbance regimes under which they evolved (Bunnell 1995; Spies and Turner 1999; Covington 2003; see chapter 2), as highlighted by examples of taxa closely associated with recently disturbed areas (e.g., Higgs and Fox 1993; Nappi et al. 2003; Buddle et al. 2006). Nevertheless, these and other species may be susceptible to novel forms and combinations of disturbances (Paine et al. 1998). For example, taxa may be maladapted to the interactive effects of two or more disturbance events in rapid succession (Paine et al. 1998), such as the compounding, cumulative, or magnified effects of intensive (and often prolonged) salvage logging following soon after a disturbance (Lindenmayer and Ough 2006).

The following example illustrates some of the potential pressures on populations inhabiting wood production forests subject to past logging and salvage logging (see also figure 3.12a). In dry years (when fires typically occur), populations must contend with (1) drought and high temperatures (which can negatively influence some species [e.g., How et al. 1984]), (2) wildfire (which can further reduce populations [Keith et al. 2002]), and (3) postfire salvage logging. Collectively, these three stressors can prove fatal to individuals, populations, and entire communities of organisms in a manner that any one of those stressors alone would not. The risks of cumulative effects arising from salvage logging and natural disturbance include major changes of ecosystem state (*sensu* May 1977) such as the conversion of rain forest to grassland as has been predicted in East Kalimantan in Indonesia (van Nieuwstadt et al. 2001).

Other kinds of cumulative effects have been hypothesized. For instance, Minshall (2003) found that fire had minor and short-term impacts on stream benthic invertebrates in the western United States. However, in burned watersheds that were subsequently salvaged, Minshall (2003) hypothesized that impacts would be significantly greater and more prolonged, particularly if more than 25 percent of the cover of merchantable timber was removed. Chan-McLeod and Bunnell (2004) suggested that extensive salvage logging, where the cumulative effects of many harvest units lead to homogenized postlogging landscapes, might make such areas more susceptible to future large-scale disturbances such as outbreaks of invertebrate pests. Nevertheless, even while we are still testing which attributes of disturbance regimes can and should be conserved in forest management (Bergeron et al. 1999; Drever et al. 2006), we have limited understanding of the cumulative impacts of multiple disturbances in rapid succession.

Any combination of disturbances that includes salvage logging is, by definition, novel because salvage logging is not a disturbance with which ecosystems or biota have evolved. Hence, any combination of disturbances that includes salvage logging has the potential to have major negative impacts on ecosystem processes and on particular elements of the biota. As an example, although tropical rain forest rarely burns, it does have the capacity to recover if subsequent high-intensity disturbances do not take place for a prolonged period (Barlow et al. 2006). This is because seed banks for many species may be activated following a wildfire, but are then exhausted if extensive mechanical harvesting follows soon after (van Nieuwstadt et al. 2001). In southeastern Asia, salvage logging of burned rain forests led to significant forest deterioration with major negative impacts on the regenerative potential of stands as well as a wide range of other undesirable effects, such as facilitating colonization by invasive grassland plants (fig. 3.12b). Similar effects have been reported for forests in northwestern North America (Roy 1956 in McIver and Starr 2000). They have also been hypothesized to occur after salvage logging of fire-damaged stands in the wet forests of Victoria, southeastern Australia (fig. 3.13b). There, seed banks for many species are activated following a wildfire, but then exhausted if extensive logging follows soon after (Lindenmayer and Ough 2006), and/or a second fire occurs (Whelan 1995), as is the case when regeneration burns are used to promote commercial crop tree germination after salvage logging.

Accelerated loss of large trees is another example of a compounding and cumulative effect of salvage logging. Large trees burned in a wildfire can either survive or, if killed, remain standing for many years during which cavity formation can occur (Inions et al. 1989). However, such trees may have a high risk of collapse if subjected to a second high-intensity fire soon after, such as a regeneration burn used to reduce log-

A **Mountain ash sequence**

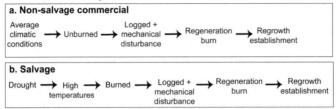

B **Rainforest disturbance sequence**

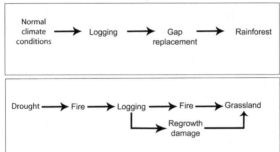

FIGURE 3.12. Comparison of simplified disturbance sequences in: A. An Australian montane ash forest, and B. A southeast Asian rainforest, following disturbance by wildfire and salvage logging.

ging slash following postfire salvage (Lindenmayer et al. 1990). Russell et al. (2006) found that snags collapsed more rapidly in salvage logged areas than in burned but unsalvaged areas. Lindenmayer et al. (1997) found similar results in the wet eucalypt forests of Victoria (southeastern Australia) and noted that in unsalvaged stands, large trees killed in a wildfire can remain standing for more than fifty years. Conversely, fire-killed trees are more likely to collapse when the surrounding stand is salvage logged and the remaining slash burned in a high-intensity fire applied to promote regeneration (Ball et al. 1999). In this case, two fires (a wildfire and a postharvest regeneration burn), in combination with logging impacts, create greater susceptibility to bole collapse than a single wildfire (Lindenmayer et al. 1990). Accelerated rates of tree loss create nesting-site shortages for an array of cavity-dependent vertebrates in wet eucalypt forests (Lindenmayer et al. 1997). Similar problems are likely to occur in forests in western North America, where up to 150 species of vertebrates rely on dead trees for nesting and denning (Rose et al. 2001; Fenger et al. 2006).

There are other important examples of cumulative effects or potential cumulative effects associated with salvage logging. After an extensive blowdown in and around Baxter State Park in Maine (USA), salvage logging was conducted outside the park. Three years later a fire burned most of the blowdown area, including both salvaged and unsalvaged areas. Soils in the salvage logged area were significantly more affected by the fire—as measured by depth of burn, residual organic soil coverage, and percentage combustible organic matter remaining—than soils in the park where salvage logging was not allowed (Hansen 1983; M. Hunter, personal communication).

Foster and Orwig (2006) hypothesized that there may be significant cumulative effects at the regional level in the northeastern United States where infestations of hemlock woolly adelgid (*Adelges tsugae*) are widespread and continuing to expand (Kizlinski et al. 2002; Orwig et al. 2002; see chapter 4). They argued that preemptive logging of eastern hemlock (*Tsuga canadensis*) stands not yet infested, coupled with both post-infestation stand mortality and salvage logging, would lead to unprecedented regionwide losses of eastern hemlock in the Northeast. Foster and Orwig (2006) noted that this prevailing pattern of extensive preemptive logging combined with both the substantial cutting of stands in decline and the additional cutting of nonsusceptible tree species was transforming "what might have been the selective decline and mortality of a single species into a regional logging activity" (Foster and Orwig 2006, 966).

Salvage Logging Effects in Landscapes with Preexisting Histories of Human Disturbance

Salvage logging generally will lead to impacts that are additive to those of green logging. Compounding effects may occur, for example, in commercially logged landscapes where key structural elements such as large, old trees are uncommon and salvage logging may exacerbate preexisting problems. Similarly, the additional road building that often accompanies salvage logging (e.g., Bunnell et al. 2004; British Columbia Ministry of Forests and Range 2006) has the potential to magnify the pre-existing impacts of roads associated with green logging (Beschta et al. 2004) as do the extensive networks of firelines created during fire suppression (Backer et

BOX 3.2
Mitigating the Effects of Fire Suppression

Attempts are made to control and suppress wildfires that occur in many forest ecosystems. Human activities associated with suppression efforts can have unintended negative consequences on ecosystems, and they may combine with salvage logging effects to magnify the overall impacts of human disturbance. Back-burning is a frequently employed wildfire suppression tactic, but these fires can burn more intensively than "natural" fires and sometimes comprise a large proportion of the overall burned area (Whelan 2002). Back-burning can also erase the landscape heterogeneity created by the inherent patchiness associated with natural fires. Firelines, roads, and firefighter camps constructed as part of suppression efforts are known to promote soil erosion and spread weeds and human refuse. These and other problems are reviewed in detail by Backer et al. (2004), and their increasing recognition has led to Minimum Impact Suppression Tactics (MIST) being employed in some jurisdictions in western North America (USDA Forest Service 2001).

al. 2004; see box 3.2). Such concerns underpinned recommendations by Beschta et al. (2004) and Karr et al. (2004) for exclusion of salvage logging from areas with limited recent human activity and, hence, where natural recovery is likely to be strong.

In the managed forests and mixed agricultural and forested landscapes of Europe, dead trees of a range of sizes and configurations have been a rare stand structural feature for a century or more. The retention of standing and fallen dead trees is an important aspect of ecosystem restoration in that part of the world (Kuuluvainen et al. 2004). Likewise, there has been extensive forest clearing for agriculture in much of the eastern United States, such that old-growth forest is primarily found on nonarable land that is usually steep, rocky, or poorly drained. Much of the region has naturally reforested after agricultural abandonment, but old-growth forest and natural disturbed forest (of all ages) are still rare (Foster and Aber 2004). Here it is important to retain both old-growth stands and naturally disturbed areas.

The Potential Magnification of Salvage Logging Impacts Arising from Postsalvage Land Management

The way salvage logged landscapes and stands are treated following logging can significantly impact post-disturbance forest recovery. Planting is widely applied following salvage logging, particularly in northwestern North America (e.g., Helvey et al. 1985; U.S. Government Accountability Office 2006). Ecological science suggests that active replanting programs can actually impede rather than assist natural ecological recovery in some forest types (Franklin 2004; Society for Conservation Biology Scientific Panel on Fire in Western U.S. Forests 2005). Poorly designed planting and revegetation programs also present risks of:

- Introducing invasive weeds (e.g., when exotic grasses are used to stabilize exposed soil) (Beschta et al. 2004)
- Replacing naturally heterogeneous landscapes with homogeneous landscapes (Crisafulli et al. 2005)
- Replacing naturally heterogeneous stands with homogeneous stands (Crisafulli et al. 2005) in which closely spaced trees lack multiple canopy layers and other features, such as asymmetrical crowns, that are characteristic of natural stands (Brokaw and Lent 1999; Franklin and van Pelt 2004)
- Elevating risks of future high-intensity wildfires by developing densely stocked stands of trees (Odion et al. 2004; Thompson et al. 2007)
- Elevating future risk of insect or fungal attack in even-aged, monotypic stands of susceptible trees (Veblen 2000; Coyle et al. 2004)

Prescriptions for post-salvage land management should consider whether accelerating rates of reforestation is actually desirable (Society for Conservation Biology Scientific Panel on Fire in Western U.S. Forests 2005).

Extensive road networks are often constructed rapidly following major natural disturbances to salvage large quantities of timber. However, roads can be major point sources of sediment (Metzeling et al. 1995), conduits for the spread of weeds (Wace 1977) and pest animals (May and Norton 1996), and avenues of access to hunters to shoot game animals (Noss and Cooperrider 1994). Beschta et al. (2004) and Bunnell et al. (2004) recommend that to limit these and other negative effects, road networks should be decommissioned soon after salvage logging.

Finally, forest recovery following salvage logging can be strongly influenced by the intensity of other disturbances, in part because some ecosystems are often not resilient to a series of perturbations in rapid succession (see chapter 2). For example, high levels of grazing by domestic livestock or overabundant native animals can impair rates of natural regeneration and slow vegetation recovery (Beschta et al. 2004; Wisdom et al. 2006). Similarly, natural patterns and rates of forest recovery can be slowed or even halted by prescribed burning soon after wildfire and subsequent salvage logging. This occurred in the Lower Cotter Catchment of the Australian Capital Territory where prescribed burning in 2006 killed young native plants that had been stimulated to germinate from soil-stored seed by recent wildfires (see chapter 4). These examples highlight a need for prescriptions for post-salvage land management to specify minimum return intervals for additional disturbances such as prescribed fire and/or to limit grazing pressure to facilitate vegetation recovery (see chapter 5).

Magnitude and Persistence of Ecological Effects

As in the case of green logging, the magnitude of ecological effects of salvage logging will vary in response to a wide range of variables:

- The quantity and kinds of biological legacies that are retained (e.g., whether an area is cleared or whether significant structural legacies are retained)
- The logging system (e.g., by ground-based machinery, elevated cable systems, or by helicopter)
- The methods used for stand regeneration (e.g., natural seeding, advanced regeneration, or artificial techniques such as planting; Lieffers et al. 2003)
- The amount and spatial pattern of logging and the proportion of a landscape that remains unharvested, either as large ecological reserves or as mid-spatial-scale protected areas (*sensu* Lindenmayer et al. 2006; see box 5.1); the location, habitat quality, and other attributes of unharvested areas, including riparian buffer strips and old-growth management areas

- The spatial extent and location of logging infrastructure such as roads, loading areas (landings), and railways
- The kinds of stands being harvested (e.g., old-growth versus secondary forest or plantations) and their ecological and other values
- The biota and ecological processes that need to be maintained in a landscape

This wide range of factors indicates that the impacts of salvage logging will be jurisdiction-specific and case-specific. In some cases, the scale and magnitude of salvage logging may exceed the size and intensity of the disturbance that preceded it (Foster and Orwig 2006).

Even when salvage logging is spatially limited, it may selectively affect parts of landscapes such as high productivity areas or the largest standing dead trees. Hence, even limited salvage logging can have large impacts on ecological processes when such kinds of sites and structures are "cherry picked."

Many of the potential ecological impacts of salvage logging outlined earlier in this chapter are overlooked or poorly understood by stakeholders. This is partly because the processes of ecosystem recovery after disturbance are still poorly understood, and partly because potential cumulative effects of natural and human disturbance have not been well documented. This oversight is important because in some cases, salvage logging impacts may have been so substantial that past interpretations of ecosystem responses to natural disturbance are incorrect and need to be reexamined. In some cases, ecosystem processes and biotic responses may have been more influenced by salvage logging than by the initial disturbance event. This appears to be true for hydrological regimes in the northeastern United States following the 1938 hurricane (Foster et al. 1997), in which the rich array of biological legacies left behind by the storm was subsequently removed. In the montane ash forests of the Central Highlands of Victoria (southeastern Australia), some of the impacts on stand structure attributed to the 1939 wildfires were at least partially related to extensive and intensive salvage logging that lasted for two decades (Lindenmayer and Ough 2006; see chapter 4). This sentiment is echoed by workers in other ecosystems, such as in the U.S. states bordering the Great Lakes (Radeloff et al. 2000), who have noted that salvage logging may affect an area larger than that of the original disturbance.

Impacts of salvage logging on biota and ecosystem processes range from short term to essentially permanent. Differences in beetle assemblages had disappeared five to seven years after fire in one study (Greenberg and Thomas 1995). Populations of carabid beetles in Canadian boreal forest actually increased in salvage logged areas compared with unsalvaged ones, most probably because of the influx of open-country species (Phillips et al. 2006). Brooks (2001) suggested that the removal of overstory insect-damaged trees would have short-term impacts on salamanders in the northeastern United States. Inbar et al. (1997) argued that the mechanical removal of logs from burned sites in a Mediterranean pine–dominated ecosystem caused an increase in soil erosion only in the first winter after fire and that it returned to normal levels in subsequent years.

In contrast to these limited and/or short-term effects, salvage logging after the 1939 wildfires in Victoria (southeastern Australia) contributed to a pronounced shortage of cavity trees for more than forty species of vertebrates—a biodiversity conservation problem that will take more than two hundred years to rectify (Lindenmayer and Ough 2006). Similarly, Beschta et al. (2004) argued that postfire logging practices in northwestern North America may have negative ecological impacts for decades to centuries.

Finally, changes in patterns of genetic variability may be permanent in cases where non-local seed sources or root stock are used in planting programs to establish vegetation on naturally disturbed and salvage logged sites. A concern that has received little or no research to date is the long-term genetic consequences of removing all trees from a disturbed forest, including those rare individuals that may have survived damage (e.g., from insects, fungus, wind, or fire).

Summary

Salvage logging impacts can be assigned to three broad and interrelated kinds of effects: altered populations of organisms, altered stand structure, and altered ecosystem processes and functions.

Salvage logging removes trees and other biological material from forests after disturbances. Such operations may reduce or eliminate biological legacies, modify rare post-disturbance habitats, influence populations, alter community composition, impair natural vegetation recovery, facilitate the colonization of invasive species, alter soil properties and nutrient levels, increase erosion, modify hydrological regimes and aquatic ecosystems, and alter patterns of landscape heterogeneity. Hence, the ecological benefits generated by large-scale disturbances (such as the creation of charred or dead trees and coarse woody debris) are diminished or eliminated by salvage logging.

Whereas most documented effects of salvage logging are negative from an ecological standpoint, others can be neutral or positive (e.g., Dobson et al. 1990), depending on the response variables measured and the landscape context. Effects are likely to vary over time and among and within vegetation types in response to the type, intensity, and periodicity of disturbance and the severity of disturbance by salvage logging.

Chapter 4

Case Studies of Salvage Logging and Its Ecological Impacts

In chapter 3 we presented a broad overview of the ecological effects that can arise from salvage logging. The aim of this chapter is to further explore these effects but in a more tangible manner through a series of short case studies. The examples contrast the different responses of biota and ecosystem processes in places where salvage logging has and has not been conducted. The chapter also provides a context for the following chapters on policies to mitigate the negative ecological impacts of salvage logging.

This chapter is divided into four sections, each addressing a different kind of disturbance—fire, insect attack, wind, and volcanic eruption. Each case study commences with a short summary about its importance and the perspectives that it contributes. The case studies are primarily those documented in the published literature.

The section on wildfires is composed of five examples, three from North America and two from Australia. The case studies span temperate forests, boreal forest, and an exotic plantation. The second section of this chapter focuses on two case studies of salvage logging following insect attack in North America, where beetle attacks are a large and rapidly increasing issue. For example, massive outbreaks of bark beetles are damaging millions of hectares of forest in western Canada as a consequence of warming winter temperatures (Carroll et al. 2004). Salvage logging often follows hurricanes, cyclones, and windstorms, and case studies from the northeastern and the southeastern United States make up the third section of this chapter.

The Mount St. Helens volcanic eruption in the U.S. state of Washington is the focus of the fourth section of this chapter. Much of the work at Mount St. Helens has underpinned new ecological thinking, particularly on the role and importance of biological legacies in shaping the trajectory of ecosystem recovery (Franklin et al. 1985, 2000; Franklin and MacMahon 2000; Dale et al. 2005).

Management objectives of an area will dictate the intensity and spatial extent of salvage logging operations, as we outlined in earlier chapters. The forests in all ten short case studies are characterized by multiple management objectives ranging from the production of large quantities of timber and pulpwood to the maintenance of an array of nonwood values, including biodiversity conservation, water production, and recreation. The primary (dominant) objective, however, varies among case studies (e.g., conservation versus water production versus timber production) and between lands of different tenure within the same case study.

Salvage Logging Following Wildfire

Yellowstone National Park, Wyoming, USA

Yellowstone National Park is the world's first gazetted national park (Clark and Minta 1994), and the primary management objectives for the area are the protection of unique geothermal features, the maintenance of biodiversity, and human recreation. Located in northeastern Wyoming, Yellowstone covers approximately 9,000 square kilometers and supports a range of environments and broad vegetation types, including grasslands and forests. Active fire suppression began sixteen years after the park was declared in 1872 (Schullery 1989), although a large number of fires occurred over the next century. Following a series of dry years, in 1988 a major conflagration burned more than 1 million hectares of the Greater Yellowstone Ecosystem (*sensu* Clark and Minta 1994) and about 250,000 hectares of Yellowstone National Park (Turner et al. 2003, fig. 4.1). The fire burned for more than three months and was finally extinguished by snowfall (Wallace et al. 2004). Many media reports described Yellowstone National Park as being "destroyed" and declared that it would "never recover" (see Wallace 2004). There were concerns not only for the forests in the park but also for the persistence of charismatic megafauna such as the grizzly bear (*Ursus arctos horribilis*), bison (*Bison bison*), and elk (*Cervus canadensis*).

BOX 4.1
Salient Features, Lessons Learned

- High fire severity
- Extensive fire size
- Public concern about ecosystem recovery
- High levels of landscape heterogeneity
- Large amounts and many types of biological legacies
- No salvage logging
- Vigorous but spatially variable postfire vegetation recovery
- Ecological surprises

Yellowstone National Park is arguably one of the most well studied environments in the world (Clark and Minta 1994), and the highly publicized 1988 wildfires have been the subject of much research (Wallace 2004). Research and other findings prove relevant to the natural disturbance and salvage logging themes that run through this book:

- Despite enormous fire-suppression efforts, it was weather conditions that ultimately extinguished the 1988 wildfires (Wallace et al. 2004).
- Although Yellowstone National Park is large, natural disturbance events such as the 1988 wildfires can be much larger and readily exceed the size of the specifically protected area.
- Yellowstone was clearly not "destroyed" by the 1988 wildfires. Rather, postfire landscapes were highly heterogeneous, reflecting marked spatial variation in fire severity, including many areas that were unburned or burned at a low intensity (Christensen et al. 1989; Turner et al. 2003). Thus, the fire left substantial numbers and types of biological legacies at both stand and landscape levels (fig. 4.2). These biological legacies subsequently had significant effects on ecosystem recovery. For example, clusters of burned fallen trees protected young regenerating trembling aspen seedlings from browsing by herbivores such as elk (Turner et al. 2003). These fallen trees have also contributed significant organic matter to otherwise nutrient-poor soils. In addition, short distances between

FIGURE 4.1. The 1988 Yellowstone wildfires (U.S. National Park Service photo by Jeff Henry).

burned and partially burned or unburned areas has given rise to "nucleated" ecosystem recovery from within the boundaries of the fire, rather than recovery being mediated through propagules sourced from outside the disturbance perimeter (Turner et al. 1997, fig. 4.3).

- Living biological legacies remaining after the 1988 wildfire included populations of grizzly bear, bison, and elk. Populations of these species fluctuated in the years following the wildfire, but all three species have persisted and some, such as the grizzly bear, have increased (Wallace 2004).

FIGURE 4.2. Spatial variation in the severity of the 1988 Yellowstone wildfires at the landscape level (U.S. National Park Service photo by Jim Peaco).

FIGURE 4.3. Young lodgepole pine (*Pinus contorta*) trees regenerating after the 1988 Yellowstone wildfires (U.S. National Park Service photo by Jim Peaco).

- Postfire vegetation recovery was vigorous and often surprising. Aspen (*Populus tremuloides*) trees, for example, established by seed rather than via clonal root sprouting in some areas (Romme et al. 1997). Other plant species changed regeneration strategy between postfire and fire-free periods (Romme et al. 1995).

In summary, Yellowstone National Park has generated many valuable new and often unexpected ecological insights into ecosystem recovery. Turner et al. (2003) noted that these insights would have been overlooked if fire-damaged stands had been salvage logged. Moreover, abundant and widespread tree and native plant regeneration occurred after the fire. This happened in the absence of any human-facilitated "restoration" programs that often follow disturbances and/or salvage logging, such as seeding to establish exotic ground covers (reviewed by Beyers 2004) or direct planting of trees (cf. Sessions et al. 2004).

Biscuit Burn, Oregon, USA

The Biscuit Burn in the state of Oregon became one of the most controversial salvage logging operations in the world (Sessions et al. 2004; Baird 2006; DellaSala et al. 2006a; Donato et al. 2006a, 2006b; Newton et al. 2006; Stokstad 2006; Thompson et al. 2007). Much of the nearly 200,000 hectares affected by the Biscuit Burn was in the Siskiyou National Forest. Management objectives for Siskiyou include wood production, watershed protection, and biodiversity conservation—many species and environments are unique to the area (DellaSala et al. 1999). The Siskiyou National Forest is within the area of the Northwest Forest Plan (Forest Ecosystem Management Assessment Team 1993; Haynes et al. 2006), which imposes an array of objectives including the maintenance of old-growth forest and conservation of forest biodiversity.

BOX 4.2
Salient Features, Lessons Learned

- Mixed (including high) fire severity
- High levels of landscape heterogeneity
- Large amounts and many types of biological legacies
- Repeated salvage logging in some areas—following the earlier (Silver) fire and the Biscuit Burn
- Vigorous but spatially variable postfire vegetation recovery
- Ground fuels initially increased by salvage logging
- Replanting after salvage logging elevated subsequent medium-term fire risk
- High levels of public controversy

The Biscuit Burn was sparked by lightning strikes in July 2002 and burned extensive areas of forest in the rugged Klamath Mountains in southern Oregon and northern California (fig. 4.4). It was the largest fire ever in Oregon and also in the continental United States outside of Alaska in 1997. Extensive salvage logging operations were proposed, and some foresters calculated that up to 5.9 million cubic meters of timber could be removed. An important feature of the Biscuit fire was that part of the area affected had burned fifteen years earlier in the Silver fire of 1987; a portion of the Silver fire had been salvage logged and planted with conifers (Thompson et al. 2007). This allowed a comparison of fire behavior on sites previously left unsalvaged and those subject to salvage logging and planting (Thompson et al. 2007). Donato et al. (2006a) compared fine fuel loads in salvaged and unsalvaged forest and the prevalence of natural regeneration on burned areas (see also Shatford et al. 2007).

The work from the Biscuit Burn had several important outcomes:

- Natural regeneration of conifers and other plant species was abundant after fire and across a wide range of dry to wet site conditions. High levels of spatial heterogeneity also characterized the natural regeneration (Shatford et al. 2007).
- Natural regeneration was slow for some species but continued to occur for up to twenty years after the Silver fire (Shatford et al. 2007).

FIGURE 4.4. Burned and salvage logged stands in the Siskiyou National Forest (photo by Dominick DellaSala).

These early postfire successional habitats recovering naturally after fire are uncommon, but they are important for some elements of the biota (see also Franklin and Agee 2003).
- There was no evidence that unsalvaged areas remained unproductive for prolonged periods (Shatford et al. 2007).
- Site history significantly influenced the severity of subsequent fires. Thompson et al. (2007) found that areas that burned in 1987 and were subsequently salvage logged and replanted were significantly more likely to reburn in 2002. These results suggested that densely spaced stands of planted conifers are susceptible to subsequent high-severity fires, even though large amounts of fire-damaged timber had been removed during salvage logging following earlier burns (Thompson et al. 2007). Thompson and his colleagues speculated that the large trees removed in salvage logging operations might not be the kind of fuel suitable for carrying wildfire. In contrast, the fine surface fuels created by salvage logging slash (see Donato et al. 2006a) and densely stocked young conifer stands planted after salvage logging might be fire-prone in the initial decades after tree establishment.

Other issues have emerged from the research and policy debates linked with the Biscuit Burn. For example, Franklin (2004, 2006) highlighted inconsistencies between intensive and extensive salvage logging operations and management objectives for the late successional reserves that were outlined by the Forest Ecosystem Management Assessment Team (1993) and then reaffirmed in the ten-year review of the Northwest Forest Plan (Haynes et al. 2006). These included the general incompatibility of salvage logging with stated goals to restore late successional or old-growth forest and to maintain populations of an array of native species, including threatened ones like the northern spotted owl (*Strix occidentalis caurina*) as previously noted by Stone (1993).

Quebec, Canada

As with other native forests, the boreal forests of the Canadian province of Quebec have multiple values—wood and pulp production, biodiversity conservation, water production, and human recreation. Management strategies aim to maintain these values.

BOX 4.3
Salient Features, Lessons Learned

- High fire severity
- High levels of landscape heterogeneity
- Large amounts and many types of biological legacies
- Nutrient cycling, vegetation recovery, and habitat values affected by salvage logging
- Salvage logging encouraged by government policies, limited provision for legacy retention

Quebec has experienced large fires in the past twenty years. Where fires occur in commercial (non-reserved) forests managed by the provincial Ministry of Natural Resources and Wildlife or by a forest products company, efforts are often made to salvage the timber within five years and before it loses economic value. Amendments to the Quebec Forest Act provide incentives for forest companies to redirect harvesting efforts, when possible, to salvage trees killed by fire, windthrow, or insects on public land (Purdon et al. 2002; Nappi et al. 2003). Extensive salvage logging occurred after large wildfires on commercial forestlands in Quebec in 1989, 1995, 1996, 2002, and 2005.

Several large fires that occurred in west-central Quebec between 1995 and 1997 were studied by researchers at the nearby Université du Québec en Abitibi-Témiscamingue (Rouyn-Noranda, Quebec) (Kafka et al. 2001; Purdon et al. 2002). In particular, a 49,070-hectare 1995 fire near the town of Lebel-sur-Quévillon and a 12,540-hectare 1997 fire near Val-Paradis were studied in detail. Burn severity mapping of the 1995 fire concluded that 3 percent of the uncut forest within the final fire perimeter survived in 43 patches ("islands") of green trees, another 32 percent of the uncut fire area consisted of patches with more green trees than dead trees, and a further 12 percent had some green trees intermixed with dead trees (Kafka et al. 2001). Despite the fact that 47 percent of the area affected by this fire had living residual structure capable of some degree of mature tree recovery and continued growth, the entire area was subject to the same prescriptions for salvage logging, which terminated further study of natural forest recovery.

Soil nutrient losses in areas subject to different burn severities and logging were quantified in another study (Brais et al. 2000). Places subject to high-severity fire were thought likely to suffer potassium depletion under normal forest rotations (65–110 years), but potassium depletion was also expected on all sites following salvage logging. Severely burned and subsequently salvage logged areas did suffer potassium, magnesium, and calcium depletion (Brais et al. 2000).

A third body of work was on the value for plant and animal biota of the young forests burned in the 1997 Val-Paradis fire. Understory cover and plant species diversity were reduced, with the composition dominated by xerophytic species, presumably because of the removal of shade provided by live and dead residual stems (Purdon et al. 2004). Evidence for significantly lower densities of regenerating conifers following salvage logging was found in another investigation (Greene et al. 2006).

Several species of wood-boring insects (families Cermbycidae and Buprestidae) were strongly associated with recently burned trees and occurred in areas characterized by high concentrations of dead trees within a landscape otherwise dominated by green trees. These invertebrates are preferred prey for several species of woodpeckers, and some researchers (e.g., Drapeau et al. 2002) argued that their reduced availability in young, naturally disturbed (but unsalvaged) forests is the factor limiting many bird species in the boreal forest. Nappi et al. (2003) reported that the black-backed woodpecker (*Picoides arcticus*) was abundant in the area burned by the 1997 Val-Paradis fire.

Current policies have no formal requirements for the retention of biological legacies in salvage-logged Quebec forests, but existing studies infer that such high-intensity salvage logging operations result in several negative effects:

- Removal of unburned patches of native forest within the boundaries of a fire
- Depletion of some important soil nutrients
- Substantially altered plant community composition, with xerophytic species being over-represented (Purdon et al. 2004) and seedbeds being extensively modified (Greene et al. 2006)
- Significant losses of critical early successional habitat for invertebrates and birds (Drapeau et al. 2002; Nappi et al. 2004)

A related issue emerging in Quebec is that early successional forests that are allowed to recover naturally following wildfire are uncommon and consequently may be of greater conservation concern than many other forest conditions, including old growth.

Central Highlands of Victoria, Australia

The montane ash forests of the Central Highlands of Victoria (southeastern Australia) contain spectacular stands of some of the tallest flowering plants in the world (fig. 4.5). Mature and old-growth trees, particularly mountain ash (*Eucalyptus regnans*) can exceed 100 meters in height and 30 meters in girth (Ashton 1975). Other key species of eucalypts in these forests include alpine ash (*E. delegatensis*) and shining gum (*E. nitens*). Montane ash forests are important for the production of timber and pulpwood (Gooday et al. 1997), water production (O'Shaughnessy and Jayasuriya 1991), and biodiversity conservation (Lindenmayer et al. 2000). Past management objectives have focused primarily on timber and pulpwood production, but more recent management goals have incorporated multiple management objectives, including timber, water, biodiversity, recreation, and aesthetic values (Commonwealth of Australia and Department of Natural Resources and Environment 1997; Victoria Department of Sustainability and Environment 2007).

The traditional form of logging in montane ash forests is clearcutting in which all merchantable trees on 15–40 hectare cutovers are harvested. After harvest, a high-intensity regeneration burn is done to promote the

BOX 4.4
Salient Features, Lessons Learned

- High fire severity
- Extensive fire size
- Large amounts and many types of biological legacies
- Prolonged salvage logging
- Depletion of key biological legacies
- Long-term effects on biodiversity

FIGURE 4.5. Old growth mountain ash forest (*Eucalyptus regnans*). The person at the bottom right highlights the size of the trees (photo by Esther Beaton).

establishment of a regrowth forest on the logged site (fig. 4.6). Forest regeneration is normally accomplished through aerial seeding of millions of *Eucalyptus* seeds collected from trees on logged sites during timber harvesting. The planned rotation time is 80–120 years (Government of Victoria 1986).

Intense wildfires are the typical form of major natural disturbance in montane ash forest, and they are often (but certainly not always) stand-replacing events (McCarthy and Lindenmayer 1998). Unlike most of Australia's 720+ species of eucalypts, those dominating montane ash forests rarely recover through epicormic growth or lignotubers if the bole is badly damaged by high severity fire. Major wildfires in 1926, 1932, 1939, and 1983 disturbed extensive areas of montane ash forest. These conflagrations were followed by salvage logging that was conducted to offset some of the economic losses created by the loss of high-quality merchantable timber (McHugh 1991). Salvage logging must occur within two years to maintain reasonable timber values due to rapid deterioration following tree death (Victoria Department of Sustainability and Environment 2003, 2007). However, profitable pulpwood logging can continue for many years following fire (Noble 1977).

Salvage logging resembles green tree clearcutting, but the disturbance order is reversed. Stands are initially burned by an unplanned wildfire.

FIGURE 4.6. Clearcutting and regeneration sequence in a stand of mountain ash (*Eucalyptus regnans*) forest (photos by David Lindenmayer).

Fire-damaged stands are then clearcut, with all merchantable timber removed (as for conventional clearcutting). The intensity of harvest and size and pattern of logged areas vary according to accessibility and fire intensity. In some cases, eucalypt regeneration is inadequate, and regeneration burns or mechanical site-preparation methods are used to reestablish eucalypt stands (see figure 1.2 in chapter 1).

The most extensive salvage logging operations in montane ash forests were conducted after the 1939 Black Friday wildfires. Salvaging continued until 1959, removing over 3.5 million cubic meters of timber from the 80-by-60-kilometer Central Highlands region alone (Noble 1977). Salvaging of fire-killed timber went on "until the deterioration in logs and the damage being done to regenerating forests called a halt" (Noble 1977, 67).

The impacts of salvage logging in montane ash forests have been examined in several recent studies (Lindenmayer and McCarthy 2002; Lindenmayer and Ough 2006), which show the following effects:

- Significant reductions in the abundance and types of biological legacies, particularly large living and dead standing trees with hollows. This, in turn, has removed critical nesting and sheltering sites essential for the persistence of cavity-using vertebrates, such as the endangered arboreal marsupial Leadbeater's Possum (*Gymnobelideus leadbeateri*).
- Reductions in the prevalence of multi-aged montane ash forests, from estimated historical background levels of 30 percent to less than 7 percent currently. Multi-aged stands are important because they typically support the highest diversity of arboreal marsupials and are key habitat for some species of forest birds.
- Reductions in vegetatively resprouting plants, such as soft tree fern (*Dicksonia antarctica*) and rough tree fern (*Cyathea australis*). Seed regenerators, which typically regenerate well after fire, also are likely to decline after salvage logging because the stimulation for germination (fire) takes place prior to mechanical disturbance (logging).
- An increase in wind-dispersed plants and those that have deep rhizomes (e.g., bracken fern [*Pteridium esculenteum*]).

The results of various observational and other kinds of studies in montane ash forests suggest that salvage logging has negative effects at a number of spatial scales. First, patterns of landscape heterogeneity are reduced as stands with the potential to become multi-aged are converted to young stands of even-aged second growth. Second, patterns of within-stand structural complexity are altered as living and dead standing trees are removed. Within these simplified stands, the composition of animal and plant assemblages is altered as functional groups (e.g., cavity-dependent vertebrates and vegetatively resprouting plants) decline or are completely lost (Lindenmayer and Ough 2006). These impacts on biodiversity have been widespread given the extensive and prolonged nature of salvage logging (Noble 1977) and have considerably exceeded the impacts of wildfires alone (Lindenmayer and Ough 2006).

Lower Cotter Catchment in the Australian Capital Territory, Australia

The Lower Cotter Catchment contributes to the water supply of the city of Canberra in the Australian Capital Territory (ACT) in southeastern Australia. Given severe droughts over the past decades in the ACT and neighboring regions, the primary management objective for the area is the production of water. The Lower Cotter Catchment was formerly partially cleared and used for grazing domestic livestock. Sedimentation and other related problems led to the catchment being revegetated many decades earlier with exotic stands of plantation radiata pine (*Pinus radiata*) that were harvested regularly for timber products.

The Lower Cotter Catchment was burned by a major wildfire in January 2003 (fig. 4.7), and extensive salvage logging operations commenced soon after to recover fire-damaged plantation timber. Ecological information on natural disturbance dynamics and the negative effects of salvage logging was ignored in initial postfire catchment management. Subsequently, a major political issue emerged related to vegetation recovery and the desirable mix of land uses given the importance of the area as a water source and the risks of extensive erosion following fire and salvage logging (White et al. 2006).

Pro-timber advocates argued that the Lower Cotter Catchment should be returned to being primarily a softwood plantation (Bartlett et al. 2005), whereas others expressed concerns about the effects of commercial plantation forestry on water quality and quantity. This debate is part of a larger worldwide controversy regarding the value and risks of plantation forests

---

BOX 4.5
Salient Features, Lessons Learned

- High fire severity
- Exotic softwood burned
- Extensive salvage logging
- Ecological surprises
- Long-term watershed issues
- Replanting after salvage logging influences catchment values
- Climate change relationships with salvage logging
- High levels of public controversy

---

in water catchments (Jackson et al. 2005). Salvage logging issues were important in these debates for several reasons but initially received little or no attention from resource managers or policymakers.

The first issue was that exotic softwood plantations are killed outright by fire and have limited opportunity for natural regeneration. Standard management prescriptions require removal of burned trees before tree planting. Following the 2003 wildfires, salvage logging of pines on burned soils contributed to major bank erosion and sedimentation in the Lower Cotter Catchment (fig. 4.7). In contrast to exotic softwood trees, eucalypt forests native to the region can recover by epicormic sprouting or other modes of recovery (Noble and Slatyer 1980; Whelan 1995), and salvage logging is not required for tree cover to be reestablished.

There were significant ecological surprises in the Lower Cotter Catchment after the 2003 wildfires. Although the area had been an exotic softwood plantation for many decades, there was natural regeneration of native overstory and understory trees—possibly from soil-stored seed banks (Adams and Attiwill 1984) and/or seed blown in from off-site. There also was an unexpected lack of severe browsing pressure by native herbivores such as brushtail possums (*Trichosurus* spp.) and the common wombat (*Vombatus ursinis*). Despite the development of natural regeneration cover, exotic grasses were sown in an attempt to stabilize the topsoil immediately following the fire and subsequent salvage logging operations. Moreover, some of the naturally regenerating native vegetation cover was cleared or treated with herbicide and the soil deep-ripped to prepare areas

FIGURE 4.7. Bank erosion and sedimentation in the Lower Cotter Catchment following the 2003 wildfires (photo by David Lindenmayer).

for pine re-establishment. Thus, the potential for native vegetation recovery was largely overlooked or ignored by resource managers and policy-makers in the Lower Cotter Catchment. The "restoration" activities adopted may have impaired native vegetation recovery. The approach taken reflects the view of foresters that conifer plantations are an appropriate protective vegetative cover for water catchments. This claim can be substantiated only when plantations are compared to agricultural or denuded lands and not to native vegetation cover.

A second issue associated with salvage logging in the Lower Cotter Catchment was consideration of possible land management and resource use scenarios beyond the salvage period. For example, climate change research suggested that the fire frequency in the ACT region will increase (Cary 2002). Shorter fire intervals mean the risk of recurrent major fires in the Lower Cotter Catchment will increase. Thus, if radiata pine stands are reestablished, there is an increased likelihood a conflagration would occur within the 30+ year rotation time typical for plantation forests in the area, necessitating the need for salvage logging to take place once again. Such problems would not occur if eucalypt forest were established because of the ability of native trees to regenerate naturally. Perhaps as important, the economics of the best kinds of land use in the Lower Cotter Catchment were highly uncertain. If the value of the water exceeds the value of the timber (assuming no change in fire intervals and no subsequent need for salvage logging), then revegetation with native trees would be the best option, especially as natural regeneration is widespread (McCarthy and Lindenmayer 2005). However, modeling work by McCarthy and Lindenmayer (2006) suggested that if native revegetation costs are large, plantation forestry may be most appropriate. These considerations clearly indicate a need for careful assessment of land-use and land-management options well beyond the initial salvage period, including the potential impacts of factors such as climate change (Spittlehouse and Stewart 2003; Lindenmayer 2006). Plantations are clearly inferior to native vegetation, including old-growth forests, as protective cover for watersheds, due to their greater vulnerability to fire and their inability to recover naturally.

In summary, salvage logging in the Lower Cotter Catchment results in:

- Reduced water catchment values; in particular, severely compromised water quality
- Loss of native vegetation that regenerated after the wildfire, as a result of damage or destruction by salvage logging operations and subsequent site preparation for replanting
- Increased invasive exotic grasses, which were sown in an attempt to stabilize salvage logged areas

The risks of future fires and consequent salvage logging if radiata pine plantations are reestablished make apparent the policy conflict between plantation timber production and the protection of watershed values. Recognition of these conflicts resulted in a reversal of an initial government decision to replant the Lower Cotter Catchment with radiata pine (Government of the ACT 2006). However, other issues remain. For example, in mid-2006, fire management authorities applied a prescribed burn to parts of the Lower Cotter Catchment to reduce levels of fuel. This damaged young seedlings that were stimulated to germinate by the wildfire just three years previously and has added significantly to the costs of programs to revegetate the Lower Cotter Catchment with native plants.

Salvage Logging Following Insect Attack

Interior British Columbia, Canada

Salvage logging of western North American lodgepole pine (*Pinus contorta* var. *latifolia*) in interior British Columbia following infestation of the mountain pine beetle (*Dendroctonus ponderosae*; fig. 4.8) is the subject of this case study. It is one of the largest salvage logging operations ever undertaken, dwarfing other salvage logging operations in Canada or the rest of the world within the last century.

The lodgepole pine forests where infestations of the mountain pine beetle are most severe are characterized by multiple management objectives ranging from the production of large quantities of timber and pulpwood to the maintenance of an array of values, including biodiversity conservation, water production, and human recreation (Hughes and Drever 2001).

As of mid-2006, approximately 9.2 million hectares of forest had been affected by infestations of the mountain pine beetle (fig. 4.9). The affected area had expanded to approximately 13 million hectares by late 2007 and is continuing to increase. High beetle population densities having spread east over the northern Rocky Mountains into the neighboring province of Alberta (Taylor et al. 2006). Although the mountain pine beetle naturally inhabits primarily mature stands of lodgepole pine (Raffa 1988; Logan and Powell 2001; Safranyik and Carroll 2006), it can feed on sufficiently large individuals of any *Pinus* species. In British Columbia, this native insect is significantly affecting western white pine (*Pinus monticola*) and ponderosa pine (*P. ponderosa*) as well as the largely noncommercial species of whitebark pine (*P. albicaulis*), all of which are rarer than lodgepole pine and often play keystone roles in some ecosystems. Infestations of the mountain pine beetle also verge on threatening extensive stands of jack pine (*P. banksiana*) in Alberta, thereby spreading eastward across the boreal forest of Canada. From the eastern boreal forest, it is possible that the mountain pine beetle could spread into a sequence of other native pine species and even down the eastern coast of North America.

Adult females of the mountain pine beetle initiate the colonization of trees and then release pheromones that attract males and additional female colonizers, resulting in aggregations of large numbers of insects in a few focal trees under endemic (non-outbreak) conditions. After mating,

> **BOX 4.6**
> *Salient Features, Lessons Learned*
>
> - Massive insect outbreak
> - Climate change relationships with natural disturbance and hence salvage logging
> - Unprecedented scale of salvage logging
> - "Bycatch" problem for undamaged forests
> - Potential ecosystem state change
> - Major forest industry readjustment

FIGURE 4.8. Mountain pine beetle (*Dendroctonus ponderosae*), adult and larval stages (photo courtesy of the Canadian Forest Service, Natural Resources Canada).

beetles lay their eggs under the bark of mature trees. Once the eggs hatch, the beetle larvae feed on the phloem tissue of the tree and disrupt nutrient flows, eventually killing the host plant. In addition, beetles carry one or more fungal associates (particularly *Ceratocystis* spp.) that disrupt the flow of water in trees and further contribute to tree death. The insect typically overwinters in the host tree as larvae, with adults emerging the next spring to disperse in search of new (living) host trees, where the cycle starts again

Cumulative Percentage of Pine Killed

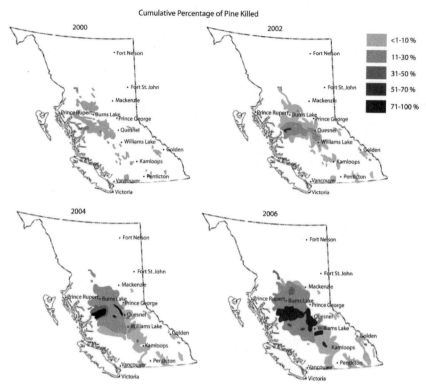

FIGURE 4.9. Maps of British Columbia, Canada, showing expanding areas of pine severely impacted by the mountain pine beetle (*Dendroctonus ponderosae*) between 2000 and 2006 (redrawn from British Columbia Ministry of Forests and Range 2006).

(Raffa 1988; Safranyik and Carroll 2006). Local or regional outbreaks of mountain pine beetle have occurred historically in interior British Columbia, the interior U.S. Pacific Northwest, and the U.S. Rocky Mountains. Many of these patches of dead forest were subsequently consumed in wildfires and played an important role in the natural forest dynamics of the region (Logan and Powell 2001; Taylor et al. 2006).

Populations of the mountain pine beetle are killed or at least kept in check by extreme winter cold (−40°C) and/or sustained low temperatures (<−25°C) in the autumn and spring. However, a series of mild winters coupled with uncharacteristically dry summers in interior British Columbia have contributed to unchecked growth of mountain pine beetle populations that have been building since 1993 or 1994 (Shore et al. 2003).

A wide range of approaches has been applied in an attempt to limit the spread of mountain pine beetle infestations. At low ("endemic" non-outbreak) population levels, the thinning of dense pine stands to improve the vigor of individual residual trees can facilitate a tree's defenses to physically extrude attacking beetles through resin production, a process known as "pitching out" (Waring and Pitman 1985; Whitehead and Russo 2005). Other methods include the use of pheromone lures, felling and burning small groups of infected trees on-site in beetle-infested pockets, and harvesting single infested stems or entire infested stands in otherwise uninfested areas (Safranyik et al. 1974; Carroll et al. 2006). An arsenic-based pesticide has been injected into individual trees to kill both the host tree and the broods of beetle larvae it supports, but there have been concerns about its persistence in the environment and its uptake by wildlife such as woodpeckers (Morrissey et al. 2007). Prescribed burns were attempted in some remote roadless areas, but the combined effects of fire and sanitation logging were limited relative to the massive land base of montane and plateau forest occupied by lodgepole pine in western Canada.

In response to the pine beetle outbreak, the British Columbia Ministry of Forests and Range has demarcated "emergency management units" (British Columbia Ministry of Forests and Range 2006) where aggressive control, containment, or post-outbreak salvage logging of affected forest stands is implemented. Salvage logging is occurring to accomplish these goals:

- Limit the spread of beetles to other trees within otherwise intact stands.
- Recover economic value; insect-damaged wood may be used for solid timber production, pulpwood, or pelletized fuelwood for several years after tree death (British Columbia Ministry of Forests and Range 2006; Byrne et al. 2006; Watson 2006).
- Sustain logging levels; timber harvesting rights on most of the public forest land in British Columbia are assigned to forest products companies. If "timber losses" due to the beetle exceed norms and are not "recovered," then allowable harvest levels (and hence revenue) must be reduced to match new estimates of resource sustainability.

- Limit the risk of fire; extensive stands of dead conifers are viewed as a serious fire hazard, first as flammable fine fuels consisting of dead foliage and small branches, and later as large fuels capable of supporting intense fires as boles collapse and fall to the ground after ten to twenty years (Fleming et al. 2002; Hawkes et al. 2005; Lynch et al. 2006; Romme et al. 2006).
- Facilitate artificial regeneration through site preparation and the planting of nursery-grown tree seedlings (British Columbia Ministry of Forests and Range 2006). Such investments are planned even in damaged stands for which the standing timber has no net commercial value (fig. 4.10).

Mountain pine beetle infestations are now so extensive that many large management initiatives are well under way. Allowable annual cut levels for public forest (set by the chief forester on behalf of the government and citizens of British Columbia) have been temporarily raised by 60 percent across the affected regions of the province (Pousette and Hawkins 2006). Existing processing facilities are running at capacity as dead wood is converted to timber, plywood, oriented strand-board, pulp and paper, despite the fact that supply would now seem to exceed demand in some markets (e.g., in the United States—the main consumer of forest products from British Columbia). In addition, new nonrenewable (i.e., unsustainable) tenures for harvesting dead pine trees (typically, those too small or too distant from mills to warrant harvesting by existing forest products companies) are being used to encourage the conversion of beetle-attacked and salvaged timber as a source of biofuel for domestic and international use (British Columbia Ministry of Forests and Range 2006).

Existing reforestation requirements (British Columbia Ministry of Forests and Range 2006) prevail for all harvested lands, with additional incentives for the conversion ("rehabilitation") of stands currently having little commercial value into conifer plantations, particularly on high-productivity areas where rates of tree growth will be most rapid (British Columbia Ministry of Forests and Range 2006). In addition, salvage logging of lodgepole pine is associated with significant "bycatch," in which many forest stands and tree species unaffected by the mountain pine beetle are being cut. Artificial forest regeneration almost exclusively uses native

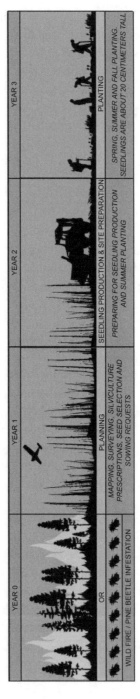

FIGURE 4.10. Schematic outline portraying part of the "Forests for Tomorrow" rehabilitation program in British Columbia, Canada (British Columbia Ministry of Forests and Range 2007). Underplanting with limited site preparation is currently preferred to minimize impact to the standing dead forest structure (graphic courtesy of the B.C. Ministry of Forests and Range).

conifers, especially lodgepole pine and interior white spruce (a natural hybrid swarm of Engelmann spruce [*Picea engelmannii*] and white spruce [*P. glauca*]). Regeneration to native hardwoods (black cottonwood [*Populus balsamifera* ssp. *trichocarpa*], trembling aspen, and paper birch [*Betula papyrifera*]) is actively discouraged during reforestation.

Many and diverse negative effects result from extensive and intensive salvage logging in interior British Columbia despite a regulated forest industry, the retention of some standing trees in most cutblocks, and environmental standards applicable to public lands (Stadt 2001; Hughes and Drever 2001; Bunnell et al. 2004). These include:

- Loss of ecologically important stand structures on logged areas, such as large dead standing trees and fallen logs critical for many elements of the biota
- Logging of unaffected spruce, fir, and hardwood trees intermixed with lodgepole pine. This is radically altering patterns of plant species composition over entire regions. In doing so, salvage logging operations are compromising short- and medium-term wildlife habitat values and midterm timber supplies (Burton 2006; Coates et al. 2006).
- Increased soil erosion, disrupted aquatic ecosystems, and increased human hunting pressure on game species through a dramatically expanded road network built to facilitate salvage logging
- Disruption of landscape-level plans for biodiversity conservation, including provision of old-growth management areas, ungulate winter ranges, and corridors of mature forest. Such disruptions are especially important to species dependent on large areas of forest interior, such as the wolverine (*Gulo gulo*), the northern goshawk (*Accipter gentiles*), and the federally listed woodland caribou (*Rangifer tarandus caribou*).
- Disruption of future timber supplies with reductions of 29 percent (from pre-uplift levels) in annual harvest expected in some areas in approximately ten years (Pedersen 2004)—a problem exacerbated by "preemptive salvage" and bycatch associated with clearcutting of mixed tree species stands.

The scale and magnitude of the mountain pine beetle outbreak is so large that it is leading to significant social and economic dislocation in the

BOX 4.7
Salient Features, Lessons Learned

- Massive insect outbreak
- Both pre- and post-disturbance salvage logging
- "Bycatch" problem in undamaged forests
- Regional changes in plant species composition

regional and national forest industries. It is also precipitating a change in ecosystem state (*sensu* May 1977; Walker et al. 2004), whereby some areas of sub-boreal forest are being converted to agriculture, primarily for grazing by domestic livestock.

Pre- and Post-insect Attack Salvage Logging in the U.S. Northeast

Infestations of hemlock woolly adelgid (*Adelges tsugae*) in the eastern hemlock (*Tsuga canadensis*) forests of the northeastern United States are the subject of a second case study of salvage logging following insect attack (Kizlinski et al. 2002; Orwig et al. 2002; Foster and Orwig 2006). Infestations of hemlock woolly adelgid may not strictly qualify as a "natural" disturbance because this is an introduced pest. Nevertheless, this case study is both interesting and important for several reasons. The eastern hemlock forests have many important values, including watershed protection (they are part of the water catchments of several large cities), habitat for many species of native plants and animals, and human recreation (Foster and Orwig 2006). Eastern hemlock is also harvested for timber and pulp, although it has low economic value (Brooks 2004; Foster and Orwig 2006).

Hemlock woolly adelgid is an aphid (fig. 4.11) that is native to Japan. It was first reported in northwestern North America in the 1920s and in the eastern U.S. state of Virginia in the 1950s (Souto et al. 1996 in Orwig et al. 2002). This introduced pest had migrated north to southern New England by the mid-1980s and has subsequently spread through much of New England, where it has killed or extensively damaged stands of eastern hemlock (fig. 4.12) in the states of Connecticut, Massachusetts, and New Hampshire (Orwig and Kittredge 2005; Preisser et al. 2007). Hemlock woolly adelgid is spread by birds, wind, and humans, making predic-

FIGURE 4.11. Hemlock woolly adelgid (*Adelges tsugae*; photo by David Orwig).

tions of spatial and temporal patterns of outbreaks difficult (Orwig et al. 2002). Moreover, it reproduces rapidly via parthenogenesis and produces two generations annually. Chemical control is not feasible in forest landscapes, but biological control agents may eventually have some success in limiting populations of hemlock woolly adelgid, although the effectiveness of controls has yet to be established (Orwig et al. 2002).

Stands of eastern hemlock have exhibited no resistance to hemlock woolly adelgid, and trees usually die within four to ten years of initial attack. Eastern hemlock stands on xeric sites are initially most susceptible to infestation once populations of hemlock woolly adelgid become abundant (Orwig et al. 2002). However, high levels of mortality eventually occur across all stands in an area (Orwig et al. 2002).

The loss of eastern hemlock trees and stands is expected to have substantial impacts on the forest ecosystems of northeastern North America (Orwig et al. 2002) because eastern hemlock is a widespread, abundant, long-lived, and shade-tolerant tree. Stands of eastern hemlock have cool, damp microclimates, a floristically depauperate understory, and slow rates of nutrient recycling with acidic litter (Rogers 1993). No other conifers can fill these combined roles in New England forests.

Two kinds of logging are taking place in the eastern hemlock stands. The first is preemptive cutting, in which hemlock trees are cut in anticipation of the future arrival of hemlock woolly adelgid (Brooks 2004). Unfortunately, tree species that are not susceptible to the adelgid are also cut

FIGURE 4.12. Extensive deterioration of eastern hemlock (*Tsuga canadensis*) following infestation by hemlock woolly adelgid (*Adelges tsugae*) at: A. the landscape level, and B. the stand level (photos by D. Lee).

at the same time to increase economic returns from harvesting because eastern hemlock is a low-value timber species (Brooks 2004; Foster and Orwig 2006). Preemptive logging is occurring well to the north of current infestations of hemlock woolly adelgid. The second form of cutting is the salvage logging of dying or dead infected stands. Like preemptive logging, this type of cutting has also been extremely widespread (Orwig et al. 2002).

Salvage logging research by a range of workers has produced important findings:

- Sites declining following infestation by hemlock woolly adelgid are characterized by the development of native understory vegetation such as black birch (*Betula lenta*) and other hardwoods that replace dying eastern hemlock trees (Kizlinksi et al. 2002; Brooks 2004).
- Compared with the effects of chronic stand decline resulting from hemlock woolly adelgid, preemptive logging and salvage logging appear to have more profound impacts on vegetation composition and ecosystem processes such as nitrogen cycling (Kizlinksi et al. 2002; Foster and Orwig 2006; Stadler et al. 2006). Thus, the impacts of salvage logging may be more substantial than the impacts of the disturbance itself (Kizlinski et al. 2002).
- Salvage logging and preemptive logging of eastern hemlock are taking place in areas that have previously been least impacted by human activities (Orwig et al. 2002), and there is a substantial bycatch issue in which many tree species that are unaffected by infestations of hemlock woolly adelgid are being harvested for timber.
- Eastern hemlock could eventually become threatened throughout its range because of salvage logging and preemptive logging. This would have corresponding additional impacts on stand regeneration, nutrient cycling, forest composition (see fig. 4.13), landscape structure and pattern, and an array of plant and animal species (Orwig et al. 2002; Brooks 2004).

Salvage Logging Following Hurricanes and Windstorms

Hurricanes, cyclones, and windstorms are major forms of disturbance in many forested ecosystems around the world. Extensive and intensive salvage logging operations have often followed these kinds of disturbances, in some cases with significant ecological impacts. This is illustrated by two case studies in this section: one on salvage logging after a hurricane in the 1930s in northeastern USA and a second on the hurricanes that have occurred in southeastern USA over the past few decades.

Post-hurricane Salvage Logging, U.S. Northeast

In 1938, a massive hurricane tracked from Long Island in New York State through the New England states of the U.S. Northeast and into Quebec, Canada. Winds exceeding 220 kilometers per hour severely damaged stands of trees across a 150-kilometer-wide strip in New England (Foster

FIGURE 4.13. Understory dominated by black birch (*Betula lenta*) after salvaged logging of eastern hemlock (*Tsuga canadensis*; photo by David Orwig).

BOX 4.8
Salient Features, Lessons Learned

- High severity
- Large amounts and many types of biological legacies
- Extensive salvage logging
- Depletion of key biological legacies
- Long-term effects on hydrological and biogeochemical processes

and Orwig 2006). The largest salvage event in U.S. history then followed, cutting more than 3.54 million cubic meters (1.5 billion board feet) of timber from an area of approximately 25,000 square kilometers (Foster et al. 1997). Large quantities of logs were stored in lakes and ponds (fig. 4.14) to prevent wood deterioration by insects and fungi until the wood could be milled.

Salvage logging had a range of significant effects. One of the most pronounced was the long-lasting shifts in hydrological regimes that were manifested at a regional scale (Foster et al. 1997). The removal of millions

FIGURE 4.14. Logs salvaged after the 1938 hurricane (photo from Harvard Forest Archives, Petersham, Massachusetts, USA).

of wind-damaged trees led to massive increases in runoff, with the Connecticut and the Merrimac rivers running much higher than normal. In addition, salvage logging altered stand structure and plant species composition over large areas, with early successional, even-aged stands of hardwoods subsequently predominating in place of uneven-aged mixed stands that were prevalent prior to the hurricane (Foster et al. 1997). Foster et al. (1997) and Foster and Orwig (2006) argue that the collective impacts of salvage logging were so substantial in magnitude and longevity that they far exceeded the ecosystem changes associated with the hurricane itself.

A seminal experiment conducted at the Harvard Forest (a Long-Term Ecological Research [LTER] site in western Massachusetts) has provided much of the scientific basis for understanding the impacts of the extensive salvage logging after the 1938 New England hurricane. The experiment replicated the effects of the 1938 hurricane by pulling down 250 trees in a seventy-five-year-old hardwood stand. Historical records from the 1938

FIGURE 4.15. Simulated hurricane experiment at Harvard Forest, western Massachusetts (photo from Harvard Forest Archives, Petersham, Massachusetts, USA).

hurricane were used to approximate the number of trees damaged, the type of damage (uprooted, snapped, bent, leaning), and the direction of tree fall (Cooper-Ellis et al. 1999). The trees pulled down in the simulated hurricane were not salvage logged but were instead closely monitored in subsequent years (fig. 4.15). An untreated control site was part of the experimental design to allow comparisons with the treated stand.

An unpredicted result was that many trees did not die, but rather survived and resprouted, producing new leaves (Foster and Boose 1995; fig. 4.16). In addition, many new tree seedlings established from seed, and considerable regeneration occurred among understory plants. Rapid vegetative recovery resulted in rainfall interception and sustained leaf transpiration, thereby limiting runoff, erosion, and nutrient losses. This was in marked contrast with the hydrological responses observed in salvage logged catchments following the 1938 hurricane (Foster et al. 1997).

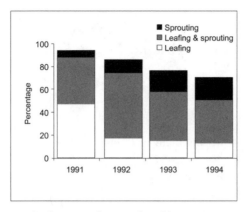

FIGURE 4.16. Survival of trees in the simulated hurricane experiment at Harvard Forest, western Massachusetts (redrawn from Foster et al. 1997).

Thus, soil microenvironments, net energy budgets, and patterns of nutrient cycling remained largely intact in the simulated hurricane experiment, despite the considerable structural changes from the manipulations (Foster et al. 1997). Moreover, the experiment highlighted the considerable capacity of ecosystems to recover naturally and rapidly from natural disturbances, even large-scale, high-intensity ones (Foster et al. 1997; Cooper-Ellis et al. 1999).

In summary, the post–1938 hurricane salvage logging operations highlighted two major kinds of salvage logging impacts:

- Significant alterations to hydrological and biogeochemical regimes at a range of spatial scales, including regional ones
- Radical modifications in stand structure and plant species composition as a consequence of the large-scale removal of the rich array of dead and living biological legacies

Post-hurricane Salvage Logging, U.S. Southeast

Our second case study of windstorm effects summarizes some of the studies of salvage logging operations that followed various hurricanes that struck parts of the southeastern United States. The U.S. Southeast supports many forest types, ranging from cypress (*Taxodium distichum*) swamps and mixed oak–hickory (*Quercus–Carya*) hardwood forests to

> ## BOX 4.9
> ### *Salient Features, Lessons Learned*
>
> - High severity
> - Climate change relationships with natural disturbance and hence salvage logging
> - Large amounts and many types of biological legacies
> - Extensive salvage logging
> - Effects on hydrological and biogeochemical processes

FIGURE 4.17. Satellite imagery of Hurricane Katrina (courtesy of National Oceanic and Atmospheric Administration, US Department of Commerce).

large expanses of native short-rotation pine plantations (especially loblolly pine, *Pinus taeda*) grown on former agricultural land. These forests and woodlands also experience insect outbreaks and wildfires as part of their disturbance regimes. However, hurricanes (e.g., fig. 4.17) can cause more damage over a larger area in a shorter period of time and therefore have a high profile in terms of perceived impacts and demands for salvage logging.

Over the last century, an average of 1.65 hurricanes per year have hit the eastern coast of the United States from Texas to Maine (Prestemon and Holmes 2004), with major hurricanes making landfall along the southeastern seaboard an average of two out of every three years (McNulty 2002). The most recent decade has seen the greatest concentration of hurricanes on record, and this activity is expected to continue or even increase (Goldenberg et al. 2001) as ocean surface waters continue to warm due to climate change (McNulty 2002; Stanturf et al. 2007).

Hurricane Hugo came ashore in September 1989, affecting more than 1.8 million hectares of forest (an estimated 50.5 million cubic meters of timber) in South Carolina and North Carolina. Subsequent field surveys and research concentrated on impact assessment (particularly in upland areas such as the Francis Marion National Forest), ways to optimize salvage logging operations, fire-hazard assessment and abatement, forest regeneration, offsetting impacts on sensitive wildlife, and the economic implications of large-scale salvage logging (e.g., Conner et al. 2004; Prestemon and Holmes 2004). Hurricane Hugo became arguably the best-studied hurricane to have reached the U.S. mainland (USDA Forest Service 1996).

Key conclusions have been drawn from the studies of Hurricane Hugo:

- Most salvage logged areas of mixed and hardwood forest were rapidly converted to short-rotation pine plantations (Conner et al. 2004). Thus, although ecosystem recovery has been rapid in naturally disturbed forests, plant species composition and other biotic responses have been significantly modified in salvage logged areas.
- Many damaged loblolly pine plantations were cleared for replanting. However, in some areas, bent and leaning pine saplings (2–4 meters tall) were staked and monitored for eight to nine years and exhibited no significant differences in growth form from untreated saplings (Gresham 2004).
- Large, weakened trees preferred by the endangered red-cockaded woodpecker (*Picoides borealis*) and other cavity nesters were disproportionately toppled by the hurricane. However, following the hurricane, various methods of initiating cavities in sound trees proved

FIGURE 4.18. Forest disturbance following Hurricane Katrina in 2005 (photo by Ricky Layson, www.bugwood.org).

successful and resulted in cavities that were used by woodpeckers. Such options were not available in areas subject to clearcut salvage logging (Hooper et al. 2004).

- Comparisons of unsalvaged and salvage logged areas between two and six years after Hurricane Hugo showed that mean soil organic matter was 48 percent greater in unsalvaged stands and mean root-to-shoot ratios of three-year-old loblolly pine seedlings were 42 percent lower (Fail 1999).
- Stream outflows from a salvage logged watershed averaged 27 percent of annual precipitation over ten years after the hurricane, compared with 22 percent of annual precipitation in an adjacent unlogged watershed. The pH of streams averaged 5.4 in the salvaged watershed, compared with 6.8 in the control watershed (Amatya et al. 2006).
- Experimental burns confirmed that fires were more intense (4.8-meter versus 0.7-meter mean flame lengths) and moved more rapidly (average of 5.5 versus 4.6 meters/minute) in hurricane debris than in debris-free areas (Wade et al. 1993). However, burning bans

and frequent patrols reduced ignitions in the early years in which debris was most flammable, after which normal prescribed burning practices were reintroduced (Saveland and Wade 1991).

- Extensive salvage logging resulted in a glut of wood fiber on regional markets, leading to a 30 percent dip in selling prices (Prestemon and Holmes 2000), although regional consumers of timber (sawmills, paper mills) benefited from these reduced prices due to the increased wood supply during salvage. Producers (woodlot and plantation owners) experienced a net transfer of wealth (as measured in terms of wood fiber values only) as damaged and salvaged stands lost value and those with the most standing timber gained value (Prestemon and Holmes 2004).

Many hurricanes have struck the southeastern United States since Hurricane Hugo (e.g., fig. 4.19), and various kinds of studies of salvage logging followed them (Stanturf et al. 2007). These investigations show that salvage logging following hurricanes can have the following results:

- Significant alterations in plant species composition and stand structural complexity
- Modified patterns of streamflow in watersheds
- Altered patterns of fire behavior

Salvage Logging Following a Volcanic Eruption

While volcanic eruptions are rarer than many other types of natural disturbances, eruptions are significant disturbance events in many regions where their scale and intensity compensate for reduced frequency. One of the most famous relatively recent examples of a volcanic eruption is that from Mount St. Helens. As outlined in the following section, the work conducted at Mount St. Helens not only produced new insights on postdisturbance ecosystem recovery but also highlighted how salvage logging can significantly alter the recovery process.

Mount St. Helens, Washington State, USA

The extremely violent eruption of Mount St. Helens, a volcano in the Cascade Range of Washington State (see fig. 4.19), on May 18, 1980,

BOX 4.10
Salient Features, Lessons Learned

- Multiple disturbance types
- Large disturbance size
- Varied pre-disturbance conditions
- High landscape heterogeneity
- Important biological legacies
- Recovery from multiple foci
- Rapid salvage and planting on private land
- Salvage logging limited on federal lands by legislation
- High biodiversity in areas excluded from salvage logging
- Ecological importance of a conifer-free post-disturbance
  community

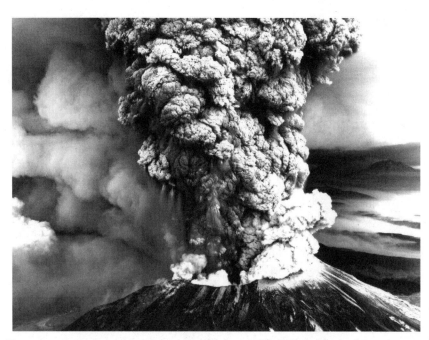

FIGURE 4.19. Mount. St. Helens erupting in 1980 (US Geological Survey / Cascades Volcano Observatory photo by Robert Krimmel).

TABLE 4.1
Broad types of disturbance at Mount St. Helens

Type	Area Affected (km²)	Thickness of Deposits (m)
Debris avalanche (landslide deposit)	60	10–195
Pyroclastic flows (glowing avalanche deposits)	15	0.25–40
Blast blowdown zone	370	0.01–1.0
Blast scorch zone	110	0.01–0.1
Mudflows (lahars)	50	0.1–10
Tephra fall (ash fall)	1,000	>0.05

initiated the large forest disturbance that is the focus of this case study. The eruption began with a collapse of the entire north face of the mountain but eventually consisted of at least six different disturbance types (Dale et al. 2005; table 4.1). Individual locations within the blast zone were affected by at least two kinds of disturbance (e.g., blast and tephra fall) and sometimes by three or four different disturbance types.

Mount St. Helens erupted within a forested region characterized by considerable diversity in local environmental conditions and vegetative cover. Mountainous topography with over 1,500 meters of relief produced variability in environmental conditions such as temperature and precipitation. The complex topography influenced the intensity and locale of many attributes of the various disturbances. Varied depths of winter snowpack still persisted at higher elevations at the time of the eruption and significantly mitigated effects of the eruption, allowing many organisms to survive. Although old-growth forests occupied much of the affected landscape, large areas had been logged during prior decades, particularly on private forestlands. Conditions on logged sites varied from recent clearcuts to dense young plantations up to forty years old.

An aggressive salvage logging program was undertaken following the eruption, beginning as quickly as safety concerns (e.g., potential for additional eruptive events) and access could be resolved. Salvage logging was conducted to capture timber values (estimates were that more than 1.5 million cubic meters of wood was available) and reduce perceived risks of wildfire and bark beetle outbreaks (Dale et al. 2005). Essentially all private and trust lands within the blast zone were salvage logged and replanted within five years of the eruption, except those traded to the federal government for inclusion in a national monument.

Salvage logging on the Gifford Pinchot National Forest proceeded more slowly because an environmental assessment was required under the National Environmental Policy Act. A decision was made to salvage log damaged timber on most of 5,600 hectares with the stated goals of thwarting insect attacks, timber decay, and future fire and to remove "timber that threatens downstream improvement" from streams. Much of the federal land was salvage logged and replanted before the plan was overtaken by the creation of the Mount St. Helens National Volcanic Monument in August 1982.

Extensive ecological research was undertaken in the post-eruption Mount St. Helens landscape and continues to the present, making it a valuable source of scientific information about disturbances, natural recovery processes, and the effects of management practices. The disturbed landscape proved to be an extraordinary laboratory because of its high diversity and spatial heterogeneity—a consequence of the nearly infinite array of environmental conditions, volcanic impacts, pre-eruption vegetative conditions, and post-eruption treatments (including salvage logging and replanting).

Many scientific findings following the Mount St. Helens eruption are relevant to the issues of natural recovery and salvage logging:

- Significant numbers of living organisms and considerable organic matter (including structures, such as snags and logs) survived the eruption throughout the blast zone (Franklin et al. 1985, 1995; Franklin and MacMahon 2000). Living legacies resulted from very diverse circumstances, including persistence in soil, lake sediments, and late-persisting snowbanks. Scientists adopted the term "biological legacies" for these organisms and organic structures (Franklin et al. 2000).
- Biological legacies were critical in both initial and subsequent recovery processes, including provision of reproductive structures or offspring, energy and nutrient sources, and protective cover and habitat.
- Biodiversity in the largely unsalvage-logged areas of the Mount St. Helens National Volcanic Monument was extraordinarily high (Dale et al. 2005; fig. 4.20). The diversity was reflected in both species richness and population levels of birds, mammals, amphibians,

and reptiles. The largely conifer-free early successional community provided important habitat for an array of vertebrates, numerous plant life-forms, and complex food webs.

- Ecological recovery was unexpectedly rapid in the blast zone, primarily as a result of the biological legacies, most notably in aquatic ecosystems (Dale et al. 2005). Furthermore, the pattern of the recovery in the blast zone was from innumerable nuclei of surviving organisms and not as a "wave front"—that is, incremental advances of invading organisms from the margins of the blast zone.

- Extensive legacies of woody debris (snags and logs) did not result in either significant wildfires or bark beetle outbreaks (Dale et al. 2005). The mantle of tephra on woody debris suppressed fire spread, and the chemical and physical properties of the volcanic ash and dead wood proved unfavorable to beetle survival and reproduction.

- Salvage logging removed critical terrestrial and aquatic habitat, primarily by removing legacies of large woody debris and snags (Dale et al. 2005). For example, it reduced habitat for cavity-dependent

FIGURE 4.20. Large numbers of biological legacies in the form of living organisms, propagules, and organic matter—much of it as coarse woody debris (logs and snags)—persisted in the Mount St. Helens blast zone despite the intensity and diversity of the disturbance processes (photo by Jerry Franklin).

birds (Crisafulli et al. 2005). It also removed large stable wood from the watercourses with negative impacts on anadromous fish (Dale et al. 2005).

• The combination of salvage logging and tree planting created tree plantations (see fig. 4.21) that have much lower levels of biodiversity as a result of the loss of key habitat components and the generation of a dense evergreen conifer layer (Crisafulli et al. 2005; Titus and Householder 2007).

In summary, studies following the Mount St. Helens eruption provided fundamental information on both natural recovery processes and the impacts of salvage logging and tree replanting. Despite the intensity of the eruption, multiple disturbance impacts, and the size of the affected region, extensive biological legacies survived and contributed substantially to re-establishment of the post-disturbance ecosystem. These legacies consisted of diverse and numerous surviving organisms, propagules, and dead organic matter, much of it in the form of woody debris (snags and logs). The woody debris had important biological and physical functions

FIGURE 4.21. Coniferous tree plantations established by planting following salvage logging have lower levels of biodiversity than areas recovering naturally with their full array of biological legacies: understory communities are sparse in the interior of this 25-year-old noble fir (*Abies procera*) plantation (photo by Jerry Franklin).

in both aquatic and terrestrial ecosystems. Reestablishment of functional ecosystems following the eruption was unexpectedly rapid because of these legacies and developed from numerous sites within the disturbed region rather than from the margins of the blast zone. Areas subject to salvage logging and the subsequent establishment of coniferous tree plantations have much lower levels of biodiversity than sites that were exempt from salvage logging and subsequent planting.

Summary

The case studies presented in this chapter highlight the fact that salvage logging is widely practiced and has been deployed after wildfires, windstorms, insect attacks, and volcanic eruptions in many parts of the world. The rationale for salvage logging; the type, extent, and intensity of operations; and the impacts of such practices on ecological processes and biodiversity vary greatly, as would be expected from the diverse set of case studies we have presented. There are some common threads to these case studies—in particular, the desire to recoup economic losses from damaged forests. Where there is human habitation, commerce, infrastructure, and movement, there is also the understandable need to remove barriers and hazards to established land uses. Another common thread is the potential for significant impacts, including some unexpected or unanticipated ones. In many cases, the impacts are commensurate with those associated with standard (green) logging operations, but the spatial scale and accelerated pace of salvage logging often accentuate the ecological consequences.

Chapter 5

Reducing the Ecological Impacts of Salvage Logging

The ecological impacts of salvage logging can be reduced using a variety of strategies, including limiting areas subject to salvage and modifying salvage practices. In this chapter we outline impact-reducing strategies based on scientific knowledge about natural disturbances and recovery processes (chapter 2), ecological effects of salvage logging (chapter 3), and lessons from well-studied cases around the world (see chapter 4). We focus on (1) identifying areas with high ecological value—at all spatial scales—where salvage logging is generally inappropriate, and (2) modifying salvage logging practices where it occurs, in order to reduce negative ecological impacts.

Importance of Management Objectives

Management objectives define the goals toward which management activities on a forest (or any other) property are directed, including the relative weights given to multiple goals. Management objectives may be predominantly economic or ecological or, in the case of many forests, a mixture of the two. Industrial forests are managed primarily for return on investment, particularly the production and harvesting of wood products. Other forests are managed primarily for biodiversity conservation or the maintenance of ecological processes (e.g., nature reserves, wildlife refuges, and municipal watersheds) or for a mixture of aesthetic, recreational, and natural values (as in most national parks and wilderness areas).

Disturbances are not expected to automatically result in changes in basic management objectives for a forest. While the management program or activities might change for a period following a major natural disturbance, the overall goals would be expected to remain—whether economic or ecological or some mix of the two. Substantial changes in goals and even ownership of property may sometimes occur after a major natural disturbance because of unique conditions created by the disturbance. For example, a substantial portion of the Mount St. Helens blast zone (Dale et al. 2005) was private, state, and federal government forest land managed primarily for wood production. After the eruption, most of these lands were acquired by the federal government and designated as a national monument because of their scientific and recreational value. But, in general, management goals for a forest property would be expected to remain much the same following a disturbance, as was the case elsewhere in the region impacted by the Mount St. Helens eruption.

Decisions regarding activities following disturbances on forests dedicated to wood production have generally followed the principle that disturbance does not alter functional objectives or goals. Hence, salvage logging of merchantable wood, site preparation, and replanting of commercially important tree species has been the typical response where timber production is a priority. Decisions regarding activities following disturbances on forests devoted primarily to values such as biodiversity conservation and the maintenance of ecological processes have often not followed that principle. Disturbances are sometimes used as an opportunity to dramatically change management objectives from ecological ones to economic ones—by implementing salvage logging where removal of timber would previously have never been allowed (see chapter 3).

This failure to consider management objectives in selecting post-disturbance activities is a major reason why salvage logging is so controversial. Salvage logging is one of the most controversial forms of forest treatment, perhaps second only to clearcutting. Debates continue about whether salvage logging reduces or increases the intensity of possible subsequent fires (e.g., Baird 2006; Donato et al. 2006a, 2006b; Thompson et al. 2007) and whether salvage logging after wildfires promotes soil compaction or helps to break up compacted soils (Beschta et al. 2004). Probably the most important underlying reason for these bitter disputes is differences among stakeholders regarding the values and uses of naturally disturbed areas (see Robinson and Zappieri 1999; Carroll et al. 2000) and the unwillingness of stakeholders from all spectrums to reference and honor the pre-disturbance management objectives for a property. Reactive management and decision making takes over, and, with it, the opportunity arises to alter pre-disturbance management goals and substitute other values and goals.

In our view, management objectives should be the starting point in any post-disturbance analysis regarding appropriate management responses, including the question of whether or not salvage logging will occur and, if so, where and how salvage operations will be conducted. Salvage logging is generally not appropriate where the primary management objectives of an area are the maintenance of key ecological processes, such as watershed protection or conservation of native biodiversity. That is, salvage logging is not consistent with the defined goals and should be either excluded or drastically limited. Conversely, where the primary management objective is wood production, broad application of salvage logging may be consistent with those objectives. That is, salvage logging is appropriate, and constraints and area exclusions should be similar to areas subject to green logging, albeit with some recognition that the terrestrial and aquatic ecosystems have already been affected by natural disturbance.

Management plans should incorporate analyses of potential disturbance events and provide at least tentative direction on what type of post-disturbance management is appropriate. This would allow all stakeholders to consider the issue of post-disturbance activities ahead of time rather than in the crisis-like atmosphere that typically follows major disturbances. Such direction is currently lacking from most forest management plans, except perhaps for simplistic and often inappropriate edicts calling for prompt erosion control and reforestation. If there are reasons to consider alterations in management objectives following the disturbance, such changes should be systematically analyzed and, in the case of public lands, vetted with stakeholders. Therefore, adjustments in management objectives should not be made in a hurried or ad hoc fashion.

Management objectives are, of course, intimately linked with issues of land tenure or ownership and access rights (Spies et al. 2007). Hence, the extent and intensity of salvage logging are likely to be different on public than on private land. However, few properties have singularly economic or ecological objectives (Keith et al. 2002); some requirements related to ecological processes and biodiversity are either self- or government-imposed. This is true even on private land where wood and fiber production is the primary goal. For example, there are requirements to incorporate at least minimal watershed standards into management of industrial plantations or fiber farms (White et al. 2006; chapter 4). More broadly, it has been argued that the integrity of ecosystem processes and biodiversity are severely compromised if land-use practices are too narrowly focused, potentially repeating mistakes in natural resource management that have been made throughout human history (Holling and Meffe 1996; Walker and Salt 2006).

Ecological considerations are important elements in most management plans. Five guiding principles underpin the maintenance of key ecological processes and forest biodiversity (Burton et al. 2006; Lindenmayer et al. 2006):

- Maintenance of connectivity for biota and ecological processes
- Maintenance of landscape heterogeneity (entailing much more than different age classes of forest)
- Maintenance of structural complexity in forest stands
- Maintenance of the integrity of aquatic ecosystems and hydrological processes
- Utilization of natural disturbance regimes as a guide for management activities (see chapter 2)

We have argued that the importance of biological legacies is probably *the* most important lesson from natural disturbance regimes that is applicable to human post-disturbance management, especially the retention of surviving (green) trees, whether as isolated individuals, in clusters or small islands, or as large landscape patches.

Multiple sources have provided comprehensive systematic guides to consideration of important ecological issues in forest landscapes (e.g., Lindenmayer and Franklin 2002; Burton et al. 2003b) (box 5.1). These guides are useful for all forests and address the maintenance of ecological values and conservation of biodiversity across multiple spatial scales — essentially from logs to landscapes. It is not possible to develop specific prescriptions to address these values that can be uncritically applied in all landscapes. This is because all ecological systems are intrinsically different as a result of their unique physical and biological properties, historical and landscape context, and current management objectives. However, it seems to us that consideration of ecological values is a universal element in responsible forest stewardship (Adamowicz and Burton 2003).

Checklists developed for "green" stands and "green" landscapes are equally applicable to naturally disturbed landscapes (box 5.1). They incorporate the multiscaled approach necessary for considering issues related to biodiversity conservation and ecological processes on forested (or other natural resource) landscapes. An initial division of a large region or domain can be into smaller areas committed primarily to the maintenance of ecological processes and/or biodiversity ("large, ecologically oriented tracts") and the remainder of the natural resource landscape, which includes lands committed primarily to economic production or to a mix of economic and ecological objectives (the "matrix") (Burton 1995; Lieffers et al. 2003).

Management of Large, Ecologically Oriented Tracts

Many large geographic regions, whether defined politically or biologically, are going to have some large forested tracts of natural vegetation (land cover) devoted primarily to the maintenance of natural ecological processes, biodiversity, or both. Reasons for these dedicated landscapes may be practical, such as goods or services (e.g., water), values (e.g., ethical or aesthetic), or a mixture of services or values (e.g., recreational sites). Examples include municipal watersheds, wildlife refuges, old-growth areas, nature reserves, wilderness, and many regional and national parks.

Salvage Logging in Large, Ecologically Oriented Tracts

Salvage logging is generally inappropriate in disturbed forests within areas exclusively or primarily committed to the maintenance of ecological processes, biodiversity conservation, or both. This generality regarding salvage logging applies whether the disturbance is considered to be of a type and intensity that is characteristic of the affected forest ecosystem or is of a novel form. However, our comment about the general inappropriateness of salvage logging on these ecologically oriented forest tracts should *not* be interpreted as a broader commentary on the appropriateness of active management in general. Active management is often an essential element of stewardship of forest resources managed primarily for ecological values, and this can certainly be the case after a major natural disturbance. Some timber removal may even be necessary to restore specific functions or infrastructure. However, traditional large-scale salvage logging activities will rarely be consistent with the ecologically oriented management objectives.

There are three primary reasons why salvage logging is generally viewed as inappropriate on large, ecologically oriented forest tracts following disturbances:

- The likelihood that such areas will exhibit high levels of natural recovery and that salvage logging will interfere with natural recovery processes
- The importance of the biological legacies to that recovery process
- Other potential negative impacts of salvage logging on ecological values, such as on the future structure and composition of disrupted forest stands and alteration of key biogeochemical and hydrological processes (Foster and Orwig 2006)

These sentiments echo those of van Nieuwstadt et al. (2001) who argue that it is important to avoid any logging of burned tropical rain forest because of effects on plant regeneration and general sensitivity of ground and understory layers in the recovery phase following fire. Similarly, Beschta et al. (2004) recommend excluding salvage logging from places with little or no recent human activity. This is because such areas are likely to exhibit strong levels of natural recovery.

Large ecological reserves (where salvage logging is excluded) are important elements of all comprehensive plans for ecologically sustainable natural resource management for at least five key reasons:

- These reserves provide some of the best examples of ecosystems, landscapes, stands, habitat, and biota and their inter-relationships as well as opportunities for natural evolutionary processes.
- Effects of human disturbance on biodiversity are poorly known, and some impacts may be irreversible. Others such as synergistic and cumulative effects can be extremely difficult to quantify or predict. These factors make large ecological reserves a valuable "safety net" relatively free from human disturbance.

- Many species find optimum conditions only within large ecological reserves, which then constitute source areas or strongholds for these species.
- Some species are intolerant of human intrusions, making it necessary to retain larger areas that are largely free of overt human manipulation.
- Large ecological reserves are "control areas" where the impacts of human activities in managed forests can be compared with natural ecosystem functions, such as the processes of forest growth, succession, and disturbance.

It can also be argued that ecological reserves subjected to large, high-intensity disturbances and from which salvage logging is excluded have added scientific value (Hutto 1995, 2006; see chapter 6). This is apparent, for example, from our case study of Yellowstone National Park (chapter 4).

Other Candidate Areas for Salvage Exclusion

There may be other large areas from which salvage logging should be excluded. Water catchments or municipal watersheds have already been mentioned (e.g., Land Conservation Council 1994). Extensive areas of old-growth forest (Lindenmayer and Ough 2006) and areas largely free of roads (Trombulak and Frissell 2000; Karr et al. 2004) are other places where the exclusion of salvage logging has been proposed. Landscapes that have a high potential for erosion, such as those dominated by steep slopes, unstable landforms, or erodible soils are other candidates for exemption from salvage logging (Karr et al. 2004).

Structurally complex (legacy-rich), naturally regenerating early successional habitats are uncommon in many landscapes (see chapter 2) and, hence, may be additional candidates for exclusion from salvage logging (Franklin and Agee 2003; Purdon et al. 2004). Key questions are the current and historical extent of such early successional habitats in a region and their importance to biota and ecological processes. In Scandinavia, a significant proportion of threatened forest species are associated with burned areas. These areas are uncommon because of a prolonged history of fire suppression (Zackrisson 1977). Consequently, forest sites disturbed by fire are rare and important in Scandinavian forest landscapes (Rülcker et al. 1994) and are strong candidates for exemption from salvage logging. Multi-aged ash-type (*Eucalyptus*) forests in the Central Highlands of Victoria (southeastern Australia) provide a second example of a rare disturbed forest condition important to biodiversity and in need of conservation. Stands of this type have been reduced from ~30 percent to 7 percent coverage in the landscape over the past seventy years (Lindenmayer and McCarthy 2002). Even in the sub-boreal and southern boreal forests of Canada—a region characterized by large disturbances (see chapter 4)—decades of fire suppression and salvage logging promoted by provincial governments have drastically reduced the area of unsalvaged, naturally regenerating young forest. All three cases indicate that past, current, and future rarity of post-disturbance vegetation types need to be considered when making decisions about salvage logging (Macdonald 2007). The newly recognized importance of naturally disturbed, unsalvaged, early successional habitat may be difficult for many stakeholders to grasp after decades during which forest activism has focused on protection of majestic old-growth forests!

Management of the Matrix

We define the matrix as semi-natural landscapes utilized for various commercial activities, including forestry and domestic grazing. In the discipline of landscape ecology, the *matrix* simply refers to the most extensive or most continuous land cover or land-use type, so it can include agricultural and urban uses in some landscapes (Forman 1995). These are landscapes where management objectives typically include a mix of both economic and ecological objectives. Hence, the matrix includes many areas where salvage logging may be an appropriate response after severe natural disturbances.

Of course, some people strongly oppose *any* salvage logging (Maser 1996) for ecological, aesthetic, or many other reasons (see, for example, the social impact study by Carroll et al. 2000). Nevertheless, the reality is that salvage logging *will* take place in many regions and landscapes—for the economic, social, and other reasons outlined in chapter 1. For example, it is inconceivable that at least some salvage logging would not occur in the millions of hectares of lodgepole pine (*Pinus contorta*) killed by the mountain pine beetle (*Dendroctonus ponderosae*) in interior British Columbia (see chapter 4). It would have been similarly inconceivable that extensive salvage operations would not have taken place in large plantation

BOX 5.1

Checklist of ecological considerations for off-reserve forest management (based on Lindenmayer et al. 2006)

Landscape-Level Conservation Strategies in Matrix Forest

Protect high ecological value areas at intermediate spatial scales

Protect special habitats
- Cliffs, caves, rockslides, etc.

Protect remnant patches of compositionally or structurally unusual forest
- Late-successional, old-growth, or early seral stages unsalvaged after natural disturbance

Protect biological hotspots
- Natural ecotones, source areas for coarse woody debris, populations of and habitats for rare species, regionally rare ecosystems

Protect disturbance refugia
- Patches that have survived past natural disturbances

Protect aquatic ecosystems and riparian buffers
- Springs, seeps, lakes, ponds, wetlands, streams, rivers, and associated buffers

Create, maintain, and protect landscape corridors and other forms of connectivity
- Natural and designed linkages of habitat types utilized by wildlife and other indigenous biota

Consider how salvage logging might exacerbate or mitigate the ecological consequences of existing landscape-level conditions

Mitigate the impacts of transportation systems
- Road, rail, and trail networks

Develop landscape-level goals for specific structural features
- Snag density, large trees with hollows, multi-layered forest

Consider the spatial and temporal patterns of timber harvesting
- Dispersed versus aggregated retention
- Size of harvest units
- Rotation lengths

Take opportunities for restoration and rehabilitation
- Re-create late-successional (old-growth) forests or other rare habitat features

BOX 5.1

Continued

Develop or maintain appropriate fire management regimes
- Maintain a range of postfire age classes and use varied prescriptions between stands

Develop management strategies for particular species
- Rare and endangered species, game species, weeds, feral animal control

Use natural disturbance regimes as a guide for logging regimes
- Time since last stand-level disturbance
- Natural disturbance refugia to exclude from logging

Stand-Level Conservation Strategies in Matrix Forest

Protect critical habitat within management units or stands

Retain structures and organisms at time of regeneration harvest
- Trees with hollows (and recruits), large decaying logs, understory thickets, and gaps

Lengthen rotation times where possible
- Designate individually retained trees and stands for management on longer rotations
- Designate variable rotations for different species or patches within stands

Consider landscape context
- Evaluate the influence of stand management activities on adjacent stands

Manage additional kinds of disturbances
- Grazing by domestic animals, recreational use, spread of invasive species

Develop targeted management strategies for particular species
- Rare and endangered species, game species, weeds, feral animal control

Protect, restore, and generate within-stand complexity

Consider natural disturbance as a guide for logging regimes
- Stand-level patterns, quantities, and variability of biological legacies

Implement appropriate fire management regimes
- Historical role of disturbance in stand dynamics and renewal

BOX 5.1
Continued

Apply novel silvicultural systems to meet stand-level goals
  • Variable retention harvest system (VRHS)
  • Harvesting with the protection of advanced regeneration
  • Novel thinning systems
Create additional structural complexity through stand management
  activities
  • Initiate snag or log formation where these features are rare
  • Create gaps in homogeneous forest
  • Promote development of multi-layered patches

estates subject to windthrow in the United Kingdom (e.g., Holtam 1971; Savill 1983) and Australia (Cremer et al. 1977) or burned plantations in Europe (Shakesby et al. 1993, 1996). There may even be situations where salvage logging is preferable to ongoing logging in undisturbed forests, if, for example, undisturbed old-growth forest is more regionally rare than disturbed forest in which salvage logging is proposed.

We begin this discussion of management of matrix lands with a brief review of ecological issues that need to be considered in all matrix forests, including those where wood production is a management objective and salvage logging is an appropriate option. Subsequently, the general management principles and the detailed checklist we presented elsewhere (Lindenmayer and Franklin 2002; Lindenmayer et al. 2006) are reconsidered specifically in the context of salvage policies and prescriptions.

Ecological Considerations for Managing Matrix Lands

A variety of stand- and landscape-level ecological issues need to be considered in managing matrix lands for economic or, most commonly, a mixture of economic and ecological goals (box 5.1) (Lindenmayer and Franklin 2002). These include issues at both landscape and patch (stand) levels.

Landscape-level considerations:

- Identification and protection or modified management of ecologically sensitive portions of the landscapes. Most prominent among these are aquatic ecosystems and hydrological networks—rivers, streams, ponds, lakes, springs, vernal pools, wetlands, and swamps and their associated riparian and hyporheic zones. Other areas of exceptional ecological value include rocky outcrops, cliffs, and rock slides; biological hotspots, such as areas with high-value spawning habitats, roosting areas for birds, camps for flying foxes, and wintering and fawning habitat for ungulates; wildlife corridors; and remnants of late successional or old-growth forest and disturbance refugia found within off-reserve forests (Mackey et al. 2002).
- Establishment of landscape-level goals (minimal or maximal amounts) of particular structures or patch types. For example, the minimum numbers of large, soft snags desired within catchments or the maximum levels of specific problematic conditions, such as the percentage of recently clearcut or prescribed burned areas within a forest landscape (Gill 1999; Parr and Andersen 2006).
- Density and condition of transportation systems (generally road networks) with regard to potential impacts on species, critical habitats, and ecological processes (Forman et al. 2002), and especially the interaction of transportation systems with aquatic systems (Diaz and Apostol 1992)
- Spatial and temporal pattern of existing harvest units or other management units as well as their relationship to other recent disturbance patches of various types and intensities
- Application and/or management of appropriate disturbance regimes, such as those involving fire (Rülcker et al. 1994) or grazing (Vera 2000)

Patch content is an important topic when assessing or planning management of both green and disturbed patches. Many stakeholders, including conservation biologists, tend to view vegetation patches in black-or-white terms—as suitable habitat or non-habitat (Lindenmayer and Fischer 2006; Lindenmayer and Hobbs 2007). In fact, patches in a forest landscape typically have varying degrees of functionality either as habitat for particular species or supporting particular ecosystem processes, depending on the structural complexity (both horizontal and vertical) encompassed within the patch. Hence, patch- or stand-level analyses are important in management, including salvage logging.

At the patch or stand levels, considerations for mitigating salvage logging impacts include the following:

- Structural retention of large hollow trees and associated recruit trees (Fries et al. 1997), understory thickets (Ough and Murphy 1998), and large fallen logs (Harmon et al. 1986; Woldendorp and Keenan 2005)
- Targeted strategies to add or create particular structures, such as girdling trees to increase quantities of dead wood (Bull and Partridge 1986) or installing nestboxes (Petty et al. 1994)
- Management of established stands to create specific structural conditions (e.g., through novel kinds of thinning [Carey et al. 1999; Douglas and Burton 2004]), regardless of their origin (e.g., planted or naturally regenerated). Most often such management will be directed at increasing structural and compositional richness within relatively uniform young stands (Lindenmayer and Franklin 2002).
- Management of open areas—such as heath and grassland—within forested landscapes that are critical for specific biota. For example, in the forests of the Swiss Jura, a reduction in the cover of trees and shrubs is considered critical for the survival of populations of the asp viper (*Vipera aspis*) (Jäggi and Baur 1999).
- Selection of return intervals for forest harvest that are consistent with recovery periods for ecological processes (Seymour and Hunter 1999) or cohort management techniques that engender a diversity of forest composition and structure within a landscape (Burton et al. 1999; Bergeron et al. 1999)
- Application of appropriate disturbance management regimes such as prescribed burning to reduce fuel loads and reduce the risk of a high-intensity fire (Covington et al. 1997; Alexander et al. 2007)

Application of Ecological Principles to Salvage Logging within the Matrix

Prescriptions for salvage logging include considerations from the patch (stand) scale through to the landscape scale—just as in management of "green" timber (Nappi et al. 2003; Beatty and Owen 2005). Thus, in this discussion, we follow the same hierarchical order as in the preceding section, moving from the landscape level (e.g., identification and management of ecologically sensitive areas at the midspatial scale) to patch- or stand-level silvicultural approaches (e.g., structural retention at the time of logging). Ecological impacts of salvage logging may be mitigated not only by prescriptions at the time of harvesting but also by appropriate planning and action before the occurrence of a natural disturbance (Macdonald 2007; Stanturf et al. 2007) and during disturbance-suppression attempts (Backer et al. 2004). We therefore include a short discussion of these topics. Although similar to the ecological approaches used to mitigate logging of "green" forest, salvage logging may require some additional considerations due to the risks of compounding disturbance impacts, such as the potential impacts of logging machinery on highly erodible burned soils (Karr et al. 2004).

The extent and intensity of salvage logging operations will be a function of management objectives and habitat rarity, as discussed earlier. Although some jurisdictions have codes of practice for salvage logging, salvage policies sensitive to biodiversity and ecological processes are often inadequate from at least some perspectives (e.g., in Quebec—Nappi et al. 2003; Saint-Germain et al. 2004; in southeastern Australia—Lindenmayer and Ough 2006; and in the northwestern United States—DellaSela et al. 2006a; Franklin 2006). For example, the environmental impacts of salvage logging do not appear to have been considered in either the British Forestry Commission response to widespread damage to Scottish forests in 1968 (Holtam 1971) or the collective response to Hurricane Gudron damage in Sweden in 2005 (Sondell 2006).

We begin consideration of the checklist presented in box 5.1 with the need to identify ecologically important areas within the landscapes broadly designated for salvage logging. By identifying critical mesoscale areas and excluding or restricting salvage logging in them, several important ecological objectives can be accomplished. The kinds of places to exempt from salvage logging include the following:

• Areas of green forest that completely or partially escaped natural disturbance and that are refugia for a broad array of surviving organisms

- Aquatic and semi-aquatic ecosystems and their adjacent terrestrial zones of influence (e.g., riparian and littoral habitat). The dead wood typically found adjacent to the aquatic ecosystems following disturbances is an important structural and material resource in these ecosystems.
- Habitats, vegetation types, and organisms otherwise poorly represented within large ecological reserves or in a landscape per se.

Maintaining Green Refugia within Naturally Disturbed Areas

Salvage logging policies need to result in the maintenance of as much of the landscape heterogeneity created by natural disturbance as possible. This is because natural forest landscapes and disturbance events are characterized by high levels of heterogeneity. Undamaged or partially damaged patches can create special conditions that are disproportionately important to ecological recovery or survival of particular biota (Mackey et al. 2002; Whelan et al. 2002).

Salvage logging of refugia can have prolonged negative ecological effects (Lindenmayer and Ough 2006). Refugia often occur in predictable and well-defined parts of landscapes. For example, the multi-aged stands of montane ash forest of southeastern Australia are most likely to be found on flat plateaus, deep south-facing slopes, and other areas with low levels of incoming radiation (Chesterfield et al. 1991; Lindenmayer et al. 1999c). As another example, while the mountain pine beetle is killing extensive stands of lodgepole pine in interior British Columbia, stands dominated by other species of trees remain essentially unaffected. Many researchers have emphasized the need to retain such undamaged areas and ensure they are not part of the bycatch when the rest of the surrounding landscape is harvested (Bunnell et al. 2004; Burton 2006; Coates et al. 2006; Griesbauer and Green 2006; see chapter 3).

Prescriptions for mesoscale salvage exclusion zones are strongly linked to the general principle of using spatial variation in the intensity of natural disturbances to determine where human disturbance may be inappropriate in a landscape or to guide the intensity of human disturbances where they take place (Morissette et al. 2002; Beatty and Owen 2005; see chapter 2). Many options exist for the identification and management of exclusion zones (box 5.2).

BOX 5.2
Deferred Harvest Areas in Western Canada

In the public forests of western Canada, several categories of mid-scale tree retention are designated to meet a variety of conservation objectives. Areas deferred from logging include the following: wildlife tree patches, retention patches, in-block structure, riparian buffers, biodiversity corridors, connectivity corridors, ungulate winter ranges, biodiversity emphasis areas, old-growth management areas, and sensitive terrain. The main feature of these mesoscale reserves is that they are not legislated as official protected areas and are not excluded from future logging. Rather, if logged, they are subject to alternative cutting practices modeled on the principles of "continuous cover forestry" (e.g., http://www.ccfg.co.uk). Typical management options include alternative logging systems (e.g., helicopter logging, selection management) or longer rotations (using group selection or irregular shelterwood systems) so that old-forest or interior forest habitat values are maintained and a stand is replaced only over a prolonged period of time (e.g., gradually over two hundred years).

Protecting Riparian Areas and Hydrological and Geomorphological Regimes

Aquatic features of forest landscapes—streams, rivers, wetlands, lakes, and ponds—are critically important to ecosystem function and biodiversity. A very large proportion of the biodiversity found in forested landscapes is associated with aquatic ecosystems, including many terrestrial as well as all aquatic organisms (Aapala et al. 1996; Soderquist and MacNally 2000; Colburn 2004). Aquatic ecosystems also play important roles in key ecological processes in addition to maintenance of hydrologic regimes, such as transport of nutrients and sediment. Salvage logging can significantly alter hydrological regimes (Foster et al. 1997; Amatya et al. 2006), and additional prescriptions may be needed to mitigate such impacts wherever possible:

- Retaining all living biological legacies to ensure that sufficient numbers of trees and understory plants continue to intercept rainfall, transpire water to the atmosphere, create microbarriers to excessive surface water flows (Foster and Orwig 2006), and maintain root strength, especially on steep, unstable slopes.

- Ensuring adequate riparian buffers are in place to protect aquatic ecosystems (Minshall 2003; Reeves et al. 2006) and provide source areas for recruitment of large woody debris; otherwise salvage logging can lead to streambank collapse, significant in-stream erosion with negative consequences of sedimentation on aquatic ecosystems (Bunnell et al. 2004), and increased water temperatures in aquatic ecosystems (Reeves et al. 2006).

- Restricting the amount of ground-based logging because it can greatly affect the physical properties of soils, including water retention, in post-disturbance environments (Beschta et al. 2004; Karr et al. 2004). Some impacts can be reduced by logging on a well-established snowpack (where available) or retaining material such as logs and logging slash on soil surfaces to limit erosion (Shakesby et al. 1993; McIver and Starr 2000). Another approach is to replace ground-based logging with cable or helicopter systems (Klock 1975; Akay et al. 2006).

- Avoiding filling water bodies with logs (fig. 5.1), as this can have severe hydrological impacts and impair the suitability of aquatic habitat for many species. This has been done in the past to protect logs from further degradation by wood-boring insects and fungi (Foster et al. 1997; Bunnell et al. 2004).

- Increasing the distance between log landings and drainage lines on steep slopes to reduce the potential for increased rates of water movement and soil erosion on burned soils (Victoria Department of Sustainability and Environment 2003, 2007)

- Planning the location and density of roads, snigging tracks (skid trails), and other infrastructure (e.g., culverts, cross drains, and water bars) to limit impacts on hydrological regimes and aquatic biota (Reeves et al. 2006)

- Removing roads once salvage logging is finished (Simon et al. 1994) because they are major point sources of sediment (Beschta 1978; O'Shaughnessy and Jayasuriya 1991) and allow access to forest areas by human hunters (Redford 1992). This may include the removal or reclamation of fire lines (firebreaks) established during firefighting operations (Minshall 2003; Backer et al. 2004).

Figure 5.1. Lake filled with logs, northeastern USA, following the 1938 Hurricane (photo courtesy of the Harvard Forest Archives, Petersham, Massachusetts, USA).

Stand-Level Management—Legacy Retention

Stand-level management can dramatically increase the contribution that salvage logging units make to the maintenance of key ecological processes and the conservation of biodiversity. In both the short and the long term, retention of live and dead wood structures will help sustain species; increase habitat suitability for particular species; improve habitat, landscape, and ecological connectivity (*sensu* Lindenmayer and Fischer 2007); buffer sensitive areas; and sustain ecosystem processes, including hydrological regimes and site productivity.

Salvage logging, by definition, removes many of the biological legacies left after natural disturbance. Evidence from a range of studies and jurisdictions suggests that many cavity-dependent species will use partially salvaged areas where some biological legacies are retained, but these species are absent from stands subject to high-intensity salvage logging (Kotliar et al. 2002; see chapter 4). Prescriptions to limit the impacts of salvaging, therefore, need to ensure that sufficient numbers of certain kinds of biological legacies are retained in salvage logged areas (Hutto and Gallo 2006; Macdonald 2007). In some cases initial retention levels may need to be significantly higher than specified under prescriptions for green logging. This is because, for example, rates of attrition and collapse of damaged trees may be accelerated by activities such as salvage logging that open up naturally disturbed stands (Lindenmayer et al. 1990; Ball et al. 1999).

Critical biological legacies for retention include live undamaged and damaged trees (Hutto 1995, 2006; Nappi et al. 2004; Hanson and North 2006) and particularly larger ones, which often have significant economic value (Morissette et al. 2002). These structures have high initial habitat value (e.g., for foraging by woodpeckers; Nappi et al. 2003; Hutto and Gallo 2006). Large living trees are important seed sources essential to the natural regeneration of disturbed areas (Greene et al. 2006). Large standing dead trees play a sequence of functional roles as they decay. Dead trees can remain standing for periods of a decade or two up to several centuries, depending on the species and environmental conditions (Gibbons and Lindenmayer 2002; Russell et al. 2006). In British Columbia, Bunnell et al. (2004) recommended the retention of tall cut stumps, or "stubs," on salvage-logged areas of lodgepole pine as cavity sites for a range of species, particularly where standing dead trees had not been retained. Similar prescriptions characterize logged forests in some parts of Scandinavia and the U.S. Pacific Northwest (Fries et al. 1997; fig. 5.2).

Large pieces of fallen timber are also important for retention. Dead wood may persist on the ground for many decades and even several centuries (Harmon et al. 1986). Large pieces of fallen timber have critical roles in hydrological and geomorphic processes, nutrient cycling and nitrogen fixation, and soil formation (Ashton 1986; Foster et al. 1997). They also are critical habitat for a wide array of biota (Harmon et al. 1986; Pharo et al. 2004; Grove and Hanula 2006).

Where management objectives include the maintenance of ecological processes and biodiversity, retention of as much of the structural and compositional legacy as possible is appropriate. However, prescriptions for the retention of biological legacies in salvaged areas are not yet well formulated in many jurisdictions (e.g., Quebec; Nappi et al. 2003, 2004). Detailed scientific studies of the relationships between levels of biological legacies and levels of ecological function are rarely available. Consequently, guidelines for retention will often have to be based on expert opinion, which can begin with consideration of the types, densities, and spatial patterns of biological legacies left following natural disturbances

A

B

FIGURE 5.2. Tall stumps or "stubs" left in a logged area in A. Sweden (photo by David Lindenmayer) and B. the Pacific Northwest of the USA (photo by Jerry Franklin).

(DeLong and Kessler 2000; Franklin and van Pelt 2004; Schmiegelow et al. 2006; see chapter 2).

Habitat and spatial requirements of particular species will also provide useful information in developing retention guidelines (Hutto 1995; Lindenmayer and McCarthy 2002; Society for Conservation Biology

Scientific Panel on Fire in Western U.S. Forests 2005). Studies of habitat requirements, foraging behavior, and prey availability for the black-backed woodpecker (*Picoides arcticus*) in northwestern Quebec revealed that large snags with limited levels of deterioration were those most appropriate for retention in burned forests targeted for salvage logging (Nappi et al. 2003). Retention may vary greatly between forests in different regions because of differences in biotic assemblages. As an example, retention levels of large cavity trees in the montane ash forests of Victoria (southeastern Australia) are different from those in the same forest types in nearby Tasmania. This is because of marked differences in the diversity and abundance of cavity-dependent species between the two regions.

Particular types of disturbed areas may need higher levels of retention of biological legacies (Hobson and Schieck 1999; Purdon et al. 2004) due to potential effects of salvage logging on many biotic elements (Kotliar et al. 2002; Lindenmayer and Noss 2006). For example, structural legacies in burned old-growth stands are likely to include very large trees and fallen logs that may be rare elsewhere in the landscape. Furthermore, large, old trees with thick bark are more likely to survive fires, which in turn means greater survival of burned trees and greater potential for development of multi-aged stands (Lindenmayer et al. 2000).

Stand-Level Management—Timing of Salvage Logging

Timing of salvage logging is a significant factor in both economic values and ecological impacts, and there are potentially many trade-offs. This is, in part, because of the rate at which degradation of timber occurs, which will vary with the tree species, disturbance type, and environmental conditions (Davies 1980; Nilsson 1974; Lewis et al. 2006). Degradation will influence the amount and the kinds of wood products that can be produced from salvaged timber, such as whether it can be used for structural wood products, pulpwood, or fuelwood.

Timing of salvage logging can also influence the ecological impacts it can have. As an example, the black-backed woodpecker in North America often uses burned stands zero to two years after a wildfire. Salvage logging after this time can have a markedly lower impact on this species, particularly if harvesting prescriptions ensure the retention of sufficient numbers of high-quality nesting trees (Hoyt and Hannon 2002). However, timing issues also need to be considered in the context of the requirements of other species that use regenerating disturbed areas but which may not be early post-disturbance specialists (Schmiegelow et al. 2006).

Timing of salvage logging may affect natural regeneration and hence the composition and structure of recovering stands (Roy 1956 in McIver and Starr 2000; Donato et al. 2006a). The most appropriate timing of salvage logging will vary markedly in response to plant life history attributes (such as the season of seed release; Greene et al. 2006) as well as other co-variables like soil type, climatic conditions, and the degree of forest floor (seedbed) disturbance. Depending on the season of a fire, for example, salvage logging that takes place rapidly after a disturbance can sometimes be completed before new seedlings and resprouters are stimulated to grow. Alternatively, rapid salvage can damage newly germinating or re-sprouting plants (Cooper-Ellis et al. 1999; Fraser et al. 2004), as observed in the rain forests of Southeast Asia (van Nieuwstadt et al. 2001). Large trees that may regenerate via mechanisms such as epicormic growth may also be removed—trees that would otherwise have survived (Hanson and North 2006) and contributed to the development of structurally complex regrowth stands.

Salvage logging that takes place rapidly after natural disturbance also can damage soil structure, which can be vulnerable to impacts of mechanical disturbance (Karr et al. 2004). In the burned plantation ecosystems of Portugal, Shakesby et al. (1993) recommended that salvage logging be delayed until leaf litter had accumulated to limit the loss of nutrient-rich ashes and soil. In the boreal forests of Quebec, Purdon et al. (2004) recommended that salvage logging take place in winter when the impacts of logging machinery on frozen soils and plants under the snow-pack would be limited (see also Macdonald 2007). However, studies in other ecosystems suggest that even over-snow logging might have negative impacts on vegetation structure and plant species composition (Stuart et al. 1993; Sexton 1998 in Beschta et al. 2004) as well as on soil microbes (Borchers and Perry 1990).

In addition to problems associated with entering stands rapidly after natural disturbance, there may be problems arising from salvage logging that continues for prolonged periods after natural disturbance (Bunnell et al. 2004). For example, prolonged salvage logging may compromise stand maturation. This is believed to have occurred in the wet eucalypt forests of southeastern Australia where salvage logging continued for two decades after wildfire (Noble 1977).

TABLE 5.1
Considerations in planning salvage logging operations

What are the management objectives for an area? What is the land tenure/ownership of the area targeted for salvage? What is the mix of values of an area?

What is the spatial context for salvage logging operations? How common or rare are naturally disturbed areas? What is the spatial extent of the naturally disturbed area? What proportion of the naturally disturbed area will be subject to salvage? How does the area of salvage compare with the area of unsalvaged forest and undisturbed forest?

What is the spatial extent and location of salvage exemption zones? Where are the areas of ecological priority and/or conservation concern?

What are the management options? What will economic returns be relative to the economic costs? Do the ecological and economic costs of cutting green (undamaged) timber exceed those of cutting trees from naturally disturbed areas?

What is the future vision for the landscape? What will the salvaged stand or landscape look like in 20, 50, 100+ years time? What might the key limiting resources be for ecological processes or organisms at those times?

What is the silvicultural system proposed for an area? What kinds and numbers of biological legacies are left by natural disturbance, and what proportion of these will be left after salvage logging?

What kinds of stands are targeted for salvage (e.g., old-growth versus regrowth) and what will their successional trajectories be if logged versus left unharvested? What will the future characteristics of these forests be, and how will these attributes relate to other forest values such as the maintenance of ecological processes and the conservation of particular elements of the biota?

Can salvage logging be conducted in a manner that the kinds, numbers, and spatial patterns of residual structures are still within the range of natural variability (or projected variability under climate change) for disturbances of this type? (Fig. 5.3)

What is the type of logging system (ground-based tractor harvesting, cable harvesting, helicopter logging, etc.), and what are its impacts on soil compaction, soil disturbance, and other attributes of forest ecosystems?

What is the length and location of the road network to access salvage blocks, and will it be permanent or will roads be debuilt and rehabilitated?

What are the protocols for post-logging site regeneration? Will they increase future fire risks, negatively alter structural features of stands, or increase risks of invasive species establishment? If planting or sowing is undertaken, what are the seed sources, and will they be from areas close to the target location for rehabilitation?

Suggestions about the need to consider the timing of salvage logging are strongly related to those about the need to appraise the ability of disturbed stands to recover naturally (Cooper-Ellis et al. 1999) and hence whether artificial restoration programs (i.e., replanting) are warranted in an area (Franklin 2004; Burton 2005; Shatford et al. 2007).

Targeted Management to Meet Particular Objectives

Targeted management practices during salvage logging operations may be needed to meet particular objectives such as maintenance or creation of particular habitat elements for species of conservation concern (Haggard and Gaines 2001). As an example, a forest management zoning and special reserve system has been established to promote the conservation of Leadbeater's Possum(*Gymnobelideus leadbeateri*) in the Central Highlands of Victoria (southeastern Australia). This zoning system is based on

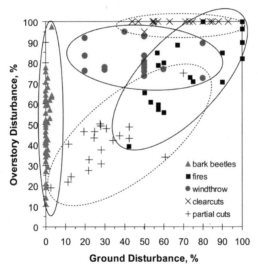

FIGURE 5.3. A reprise of disturbance classification, with understory and soil disturbance axes (often highly correlated) combined and clearcut and partial-cut logging samples added to the datasets portrayed in Fig. 2.12. Such assessments, or model projections, of the range of natural variability in the structural diversity of disturbed stands provide guidance for the level of salvage logging that probably can be sustained in an ecosystem, indicated here by overlaps between solid and dashed ellipses (P. Burton, unpublished data).

well-known habitat relationships for the species (Macfarlane et al. 1998). Salvage logging is not permitted in Leadbeater's Possum management units.

Summary

A range of considerations is relevant when planning salvage logging (table 5.1). Where land management objectives involve multiple values (which is generally the case), policies and prescriptions specific to a given land-scape or particular set of conditions can be broadly guided by the same general principles applicable to green logging. These principles are aimed at maintaining critical ecological processes and biodiversity across the range of spatial scales—that is, regions, landscapes, stands, and structural complexity within stands. Some general guidance on these four broad principles comes from the use of natural disturbances as a template (e.g., fig. 5.3) to inform the location, intensity, and frequency of human disturbance regimes (in this case, salvage logging).

Chapter 6

Toward Better Management of Naturally Disturbed Forests

There is a need for broader dialogue about the management of forests that experience major natural disturbances, particularly as we enter a century likely to experience disturbances that are at scales, intensities, and forms that have no historical precedent. This chapter is a contribution to this dialogue, with comments on three main points. First, there is clear necessity for developing long-term visions for landscapes subject to natural disturbances, which are often then subjected to salvage logging. These long-term visions are needed to foster planning and actions that are proactive rather than reactive. Second, we cannot overstate the importance of adaptive management, including long-term monitoring of salvage logged and unlogged stands, additional research on disturbances and recovery processes, and designed studies of ecological impacts of salvage logging and other post-disturbance management practices. And, third, the language used in discussing natural disturbances and human responses to those disturbances is critically important, because value-laden language often colors discussions about how ecosystems respond to disturbances.

The Psyche of Management—Avoiding Crisis Management

Large-scale natural disturbances commonly lead to great stress on personnel within agencies charged with the management of natural resources (Cary et al. 2003). Fighting fires, responding to large windstorms and hurricanes demand rapid and substantive actions. However, a crisis management psyche can continue even after natural disturbances have passed, leading to hasty decision making (Robinson and Zappieri 1999) and errors in management practices on the ground (DelaSella et al. 2006a). This can have negative environmental outcomes, sometimes for prolonged periods (Foster and Orwig 2006; Lindenmayer and Noss 2006).

Calls by the media for rapid responses such as "cleaning up the mess" and by politicians for "disaster relief" after major natural disturbances contribute to hasty and poor decision making. This was evident from salvage logging operations that followed the windstorm that occurred in southeastern England in October 1987:

> [T]here was an immediate sense of urgency, stoked up by the press. Action was a substitute for thought. All through that [following] very wet winter, machines galumped through the woods, getting out timber which was sold at bottom prices. . . . Ecological damage done by clearing up and replanting exceeded that done by the storm itself. (Rackham 2001, 202)

Decision making during crisis management has a poor record in environmental science, such as in areas of weed control (McNeeley et al. 2003) and threatened-species conservation (Groom et al. 2005). Salvage logging is no exception, because the importance of other forest values and long-term ecological sustainability can be overlooked. As outlined in previous chapters, non-timber values such as water quality and quantity, biodiversity conservation, and ecosystem resilience (sensu Folke et al. 2004; Walker et al. 2004) need careful appraisal (see chapters 3 and 4).

Predisturbance Planning

It is better to anticipate extreme events and prepare contingency plans before they occur, rather than allow events to drive management responses. Although the timing and location of major disturbances can seldom be known in advance, resource management agencies need to plan how they will respond to them, including protocols related to salvage logging and reforestation (Stanturf et al. 2007). In some cases it will be appropriate to recognize that large-scale disturbances are inevitable and to adjust sustained-yield calculations accordingly. This will allow expected timber losses to be factored into future estimates of timber flows and limit risks of significant overcutting when major natural disturbances take place. There is already an extensive body of work on how reserves can be designed so that the impacts of disturbance regimes can be incorporated (e.g., Baker 1992), and similar kinds of thinking need to be extended to off-reserve areas (Lindenmayer and Franklin 2002). Resource managers and policymakers also need to be cognizant that, in many cases, it is a sequence of disturbances (or even the disturbance regime per se) rather than a single event that will have the most profound influences on ecosystems (Gill 1975).

Pre-disturbance planning should include (1) identifying the types and locations of areas that will not be salvage logged, (2) the level of legacy retention in areas where salvage logging will take place, and (3) protocols for other kinds of treatments following salvage operations (see chapter 5). For example, Lindenmayer and Ough (2006) recommend that existing areas of old-growth montane ash forest in the Central Highlands of Victoria (southeastern Australia) be excluded from salvage logging (box 6.1) for the following reasons:

- Old-growth forest is uncommon in wood production zones.
- Large-diameter trees characteristic of old-growth forest are likely to survive fire, allowing burned stands to develop into multi-aged forests (Lindenmayer et al. 2000).
- Multi-aged stands created following old-growth burns will have special habitat values for a range of mammals and birds.
- Increasing the area of multi-aged forest will increase levels of congruence between spatial patterns found in wood production landscapes and natural (unlogged) landscapes.
- Large-diameter trees characteristic of old-growth forest will stand longer than small-diameter trees and therefore have a greater chance of providing suitable cavity trees for prolonged periods (Lindenmayer et al. 1997).
- Large-diameter trees characteristic of old-growth forest will be important substrates for the development of luxuriant mats of bryophytes (Ashton 1986; Lindenmayer et al. 1999b), which may act as micro-firebreaks in future wildfires (see Andrew et al. 2000).
- Stands of old-growth montane ash forest have been mapped in the Central Highlands of Victoria and many occur in "Special Protection Zones," making it possible to preplan and map areas exempted from salvage logging.

BOX 6.1

Planning in Advance for Salvage Logging — The Case of Victorian Ash–Type Eucalypt Forests, Southeastern Australia

Large wildfires are the primary kind of natural disturbance in Victorian montane ash forests (Ashton 1981; Mackey et al. 2002). The question is not whether, but when, montane ash forests will burn again. The average fire return interval in montane ash forests is approximately 110 years (McCarthy et al. 1999), but recent simulation modeling suggests that the frequency of intense large-scale fires will increase (Cary 2002) in response to rising temperatures and more extended droughts. As a consequence, it is critical for management agencies to be planning future salvage logging operations well in advance. As discussed in chapter 4, past salvage logging operations have had prolonged negative impacts on montane ash forests (see also Lindenmayer et al. 2004b; Lindenmayer and Ough 2006). Therefore, spatial variation in the intensity of salvage logging will be essential, mimicking some of the variability found in wildfires. Zones from which salvage logging is excluded will be important. Salvage logging is already precluded from the ~20 percent of the montane ash forest resource that occurs in the Yarra Ranges National Park (Land Conservation Council 1994). Outside this important reserved area, existing limited areas of old-growth forest, if burned, should also be exempt from salvaging because of the likelihood that either multi-aged forests will be created or large dead trees with hollows will form and become critical nest sites for many cavity-dependent species (Lindenmayer and Ough 2006). Gullies and other areas that are important for the protection of waterways and catchment values should also feature among salvage exclusion zones.

Codes of Practice have been developed for conventional logging operations (Victoria Department of Natural Resources and Environment 1996), and many of the prescriptions, such as snag retention, are also relevant for areas subject to salvage logging. However, additional prescriptions are warranted and should be developed well before the occurrence of the next major conflagration. As one of many examples, harvesting should be excluded from small areas of unburned forest that occur within the boundaries of planned cutover units.

Given the present limited understanding of salvage logging effects in montane ash forests (Lindenmayer and Ough 2006), carefully planned field experiments have already been designed. These experiments will be implemented when the next large-scale wildfire occurs.

Scenario planning (e.g., Santelmann et al. 2004) and decision support tools (e.g., Angst and Volz 2002) can be useful for pre-planning responses to major events such as large-scale natural disturbances and can facilitate the explicit consideration of alternative futures (Peterson et al. 2003). Such planning will become increasingly important if forecasts of more frequent and/or more intensive large-scale natural disturbances are accurate. Resource managers would be wise to expect such disturbances, plan responses well in advance, and account for their effects in calculations of sustained yields and rotation lengths (e.g., see Kohnle and von Teuffel 2004).

The potential for novel and unanticipated disturbances means that it may not always be possible to instigate adequate and/or appropriate pre-disturbance planning. As an example, neither the spatial extent of the mountain pine beetle (*Dendroctonus ponderosae*) outbreak in the sub-boreal forests of British Columbia nor the magnitude of salvage logging operations following the outbreak could have been anticipated (see chapter 4). In fact, the highly altered landscapes created by prolonged and widespread salvage logging are unlikely to support long-term viability of the timber industry (at pre-outbreak levels); huge shifts are likely in the types and amounts of forest goods as well as services (including wood) they are capable of producing (see chapter 4). These kinds of social and economic issues are just beginning to receive the attention of stakeholders, including resource managers and policymakers responsible for sustaining ecologically, socially, and economically productive timber industries. In some areas of inland British Columbia, provincial government discussions include the possibility that a large portion of the forest products industry will transition from timber and pulpwood to fuelwood and biofuel (British Columbia Ministry of Forests and Range 2006). Maintenance of key ecological processes (e.g., watershed processes and values) and biodiversity will nevertheless need to be a fundamental part of landscape and regional planning in the British Columbia interior.

Limiting the Need for Salvage Logging — Pre-disturbance Management

The need to consider instigating salvage logging may be avoided in the first place by management activities that reduce either the risks of major natural disturbances or the intensity of such disturbances. The appropriateness of these practices will vary among forest types (Brown et al. 2004).

For example, lower levels of stand thinning, closer spacing of non-salvage cutover units, and careful location of boundaries of cutover units may limit the potential for extensive windthrow in some forests (Franklin and Forman 1987; Cannell and Coutts 1988; Rowan et al. 2002) but may increase windthrow risks in other kinds of forests (Cremer et al. 1977). It has been suggested that created stands with mixed structures and composition can reduce vulnerability to insect attack (Miller and Rusnock 1993). Stand thinning has been proposed as a way to limit future pest infestations in other jurisdictions (Whitehead and Russo 2005), but such an approach has had mixed success and in some cases may even have led to increased subsequent impacts (Muzika and Liebold 2000). Similarly, consideration of the spacing of trees during post-salvage site rehabilitation, together with the establishment of strips with lower levels of fuel, may reduce the risks of future stand-replacing fires in areas such as the Douglas-fir (*Pseudotsuga menziesii*) forests of the U.S. Pacific Northwest (Franklin 2005; Thompson et al. 2007). Stand-thinning methods have been applied to limit the risks of high-intensity wildfires in the ponderosa pine (*Pinus ponderosa*) forests of the southwestern United States (Noss et al. 2006) as well as stands of other tree species that grow in association with ponderosa pine (Brown et al. 2004). These methods are used because prolonged fire suppression and other management activities have altered fire regimes in these forests from a frequent low-severity regime to an uncharacteristic high-severity one (Spies et al. 2004; Noss et al. 2006).

Prescribed burning of mature forests is an increasingly used management technique designed to limit the risks of high-intensity wildfires (Gill et al. 1999; Converse et al. 2006; Alexander et al. 2007). Controlling prescribed fires depends on topography, access, wind speed, temperature, humidity, fuel loads, and fuel moisture content (Whelan 1995). Prescribed burning can be controversial and may have negative effects on some biota and ecological processes (Syphard et al. 2006). Some vegetation types and the biota they support are susceptible to any fire (Bradstock et al. 2002); in others, frequent burning can select for fire-prone species, making some vegetation types more flammable than they might otherwise have been. Prescribed burning will not be appropriate in forest types such as the cool, wet ash forests and rain forests of the Australian states of Victoria and Tasmania, the coastal or interior rain forests of the Pacific Northwest and British Columbia, or the tropical rain forests (Barlow et al. 2006), where luxuriant ground and understory layers mean that only high-intensity wildfires are possible under unusually dry conditions.

Prescribed burning may well be appropriate where prolonged periods of active fire suppression have changed fire regimes from recurrent low-intensity fires to uncharacteristic high-intensity, stand-replacing wildfires (Spies et al. 2004; Noss et al. 2006; Donovan and Brown 2007). Its use may help restore more "natural" historical disturbance regimes in some vegetation types (e.g., Zackrisson 1977; Kotliar et al. 2007), thereby creating appropriate habitats for some taxa (Hyvärinen et al. 2006) and limiting the risks of major conflagrations that can be stand-replacing events (Brown et al. 2004).

Some authors have called for other kinds of pre-disturbance management activities to limit the impacts of salvage logging. For example, Karr et al. (2004) suggested that barriers to the movement of aquatic organisms such as fish should be removed from many watersheds to both speed the post-disturbance population recovery and increase resilience to natural and human disturbances. Similarly, the behavior and intensity of wildfires are typically spatially heterogeneous (Agee 1993; Turner et al. 2003; Burton et al. 2008), and unburned or partially burned patches can be crucial for the post-disturbance recovery of many organisms as well as the maintenance of some ecological processes (Franklin et al. 2000; Whelan et al. 2002). Given this inherent variability in forest fires, one approach to limiting the potential impacts of salvage logging is to ensure that back-burning during efforts to control wildfires does not lead to the loss of all unburned areas (Backer et al. 2004). Of course, unburned or partially burned areas might be among the prime candidates for exemption from logging once salvage operations commence.

Preemptive Salvage Logging and Forest Resilience

Preemptive salvage logging is a form of harvesting sometimes used in the belief that it will (1) confer greater resilience to forests against future disturbances (Kyker-Snowman 2000), and (2) reduce the impacts of such future disturbances. However, little or no empirical data to support these beliefs currently exist. Conversely, such management activities may have greater negative impacts on forest ecosystems and key ecosystem processes than leaving naturally disturbed forests free of human intervention (Foster and Orwig 2006).

A special case of pre-disturbance salvage logging is occurring in the forests of the northeastern United States where infestations of hemlock woolly adelgid (*Adelges tsugae*) are causing widespread mortality among eastern hemlock (*Tsuga canadensis*) stands (see chapter 4). Eastern hemlock trees are often being logged before the arrival of hemlock woolly adelgid, based on the assumption that they will inevitably become infested, foster future population growth of the insect, and die once the aphid arrives. Other rationales for salvage logging of eastern hemlock include public safety, protection of water catchments from soil erosion and excessive nutrient loads, and economic returns from harvested trees. Because eastern hemlock is a low-value timber tree, other unaffected tree species are often cut at the same time to increase economic returns (Brooks 2004). Thus, pre-infestation logging of eastern hemlock and cutting of non-susceptible trees is leading to major regionwide effects on forest composition and biodiversity (Foster and Orwig 2006; see chapter 4).

The Need for Additional Research

New knowledge to inform policies and on-the-ground operations is a crucially important part of improving the management of salvage logging. Policies on salvage logging need to consider not only the ecological factors primarily discussed throughout this book but also the economic and long-term social welfare issues — not just wood production and biodiversity conservation objectives. We offer general recommendations for conserving what we consider to be valuable features, such as the structural complexity of recovering stands, and the potential to maintain ecosystem processes and services as well as create habitat for particular species. But we know little about thresholds of biodiversity response to the ecological factors we are capable of manipulating in forests (Groffman et al. 2006; Walker and Salt 2006). In other cases, there are currently limited empirical data to support postulates that, for example, windthrown timber is a major fire hazard (Kulakowski and Veblen 2007) or that "forest protection silviculture" will confer greater resilience to future disturbances (Foster and Orwig 2006).

Learning from Large-scale Natural Disturbances

Large-scale disturbances provide important opportunities to learn about disturbed ecosystems and their recovery, thereby offering a chance to improve future policies and practices for natural resource management. For example, critical new and surprising insights on forest dynamics were gained from studies of the impacts and recovery processes associated with the volcanic eruption of Mount St. Helens in the U.S. Pacific Northwest (Franklin et al. 1985; Franklin and McMahon 2000) and with the large wildfires in Yellowstone National Park (Turner et al. 2003; see chapter 4).

Focused studies on the ecological effects of salvage logging operations are relatively rare (McIver and Starr 2000; Noss and Lindenmayer 2006), and much more knowledge is needed to inform natural resource management. An important question to address is, How can we ensure that the recovery of disturbed ecosystems is not impaired? Such research can help resource managers avoid repeating past mistakes.

Pre-disturbance planning for research projects and the acquisition of necessary infrastructure (such as mobile laboratories) is important to maximizing opportunities to study natural disturbances and recovery processes. It allows for new approaches to observation, management, and experimentation with naturally disturbed areas to be conceived in advance, providing templates that can be adopted when large disturbances occur.

The Role of Experiments and Cross-sectional/Retrospective Studies

Using a variety of management prescriptions and research approaches can maximize learning opportunities where salvage logging occurs. Research approaches can include experiments, natural experiments, and observational studies. Careful design can avoid deficiencies associated with much past research on salvage logging. Some issues with previous research include limitations on inference related to experimental design (see McIver and Starr 2000). These include:

- A lack of disturbed but unsalvaged sites (Greenberg and McGrane 1996; Elliott et al. 2002; Khetmalas et al. 2002).
- Absence of pre-disturbance data for salvaged sites (Greenberg et al. 1994).
- Lack of control sites where no disturbance has occurred.
- Limited replication of treatment and control sites (Blake 1983; Cooper-Ellis et al. 1999; Phillips et al. 2006; Saab et al. 2007).
- Limited quantification of the severity of the natural disturbance (Smucker et al. 2005; Kotliar et al. 2007) either to document conditions immediately prior to salvage logging or to act as "control" sites for areas that are salvage logged.

FIGURE 6.1. Experimental burning of insect-damaged stands in interior British Columbia (British Columbia Forest Service photo by Bryan Bowman).

Experiments are the most rigorous form of ecological research. Perhaps the most celebrated forest disturbance experiment is one initiated by David Foster and his colleagues (1997; Cooper-Ellis et al. 1999) that simulated hurricane disturbance of forests in the U.S. state of Massachusetts (see chapter 4). Salvage logging impacts on stand regeneration and plant species composition are being studied in the Canadian province of Quebec (Purdon et al. 2004; Greene et al. 2006). The Carrot Lake Experimental Fire Study (fig. 6.1) is a new experiment in British Columbia that is quantifying relationships between insect-damaged stands and fire behavior, in support of improving both fire-protection activities and the use of prescribed fire for the renewal of lodgepole pine (*Pinus contorta*) stands as an alternative to salvage logging.

It is not always possible to establish true experiments or take advantage of natural experiments due to logistical, financial, and other reasons (McIver and McNeil 2006). The small spatial scale of many experiments, and the high degree of control needed over many potentially important factors, can limit inferences such as those for wide-ranging mammals and birds (Wiens et al. 1997). Therefore, improved understanding of salvage logging can also be gained from several kinds of non-experimental approaches, such as retrospective, observational, chronosequence, and modelling studies. Simulation modelling can assist the understanding of variations in fire severity over large spatial scales and prolonged time frames (Thompson et al. 2007). Retrospective and chronosequence studies are useful in forest ecosystems where trees can be long-lived and where much of the research depends on substituting space for time (i.e., stands at different stages since disturbance are compared). However, these approaches require care because results can be confounded by site differences and past histories (Pickett 1989); also, records on past histories, such as silvicultural treatments, are often lacking.

Because retrospective and observational studies will continue to be a significant part of forest research, there is a need to assess and carefully document the following:

- The characteristics of stands following natural disturbance, such as damage classes (e.g., Smith and Woodgate 1985; Kotliar et al. 2007).
- Levels of heterogeneity within the boundaries of a disturbance and, for example, the amount and spatial variability in living and dead biological legacies (Delong and Kessler 2000; Schmiegelow et al. 2006)
- The nature of salvage logging by recording, in particular, the cutting method and the kinds, sizes, and densities of retained structural features (Foster et al. 1997). These records facilitate the use of managed sites in retrospective studies, investigating, for example, the effects of regeneration on biodiversity.

Wherever possible, it is useful to leave unsalvaged areas of at least several hectares in size on different site types. They can provide control areas for comparative analysis and opportunities for future research in addition to helping to sustain biodiversity values and preserve some key ecological processes.

Documenting the location, size, and type of salvage operations is also important for historical purposes. Indeed, the writing of this book was made more difficult because of limited information on the extent of burned, windthrown, or insect-damaged areas and of the degree to which those disturbed areas were salvage logged.

Integrating Research, Policy, and Management—
Adaptive Management

Results of research on salvage logging need to be utilized in developing improved policies and management practices (Russell-Smith et al. 2003). Hence, links among policymaking, operational practices, and salvage logging research are critical. Active adaptive management is the most highly discussed framework for connecting policy, management prescriptions, and research in natural resource management (see Holling 1978; Walters 1986; Walters and Holling 1990). A detailed discussion of adaptive management is beyond the scope of this book but can be found elsewhere (e.g., Lee 1999; Lindenmayer and Franklin 2002; Bunnell et al. 2003). Briefly, active adaptive management works as follows: Results from experience or research can underpin improvements in salvage logging policies, which, in turn, lead to improved management prescriptions for salvage logging. New prescriptions are, in turn, rigorously assessed in new salvage logging experiments or operational trials, with the results being fed back to policymakers and managers—thereby closing the cycle of knowledge gathering, policy development, and improved management application (fig. 6.2).

There are very few applications of active adaptive management in forest management anywhere in the world, despite an extensive literature on the subject (Simberloff 1998; Stankey et al. 2003). Here are some of the reasons (after Davis et al. 2001; Lindenmayer and Franklin 2002):

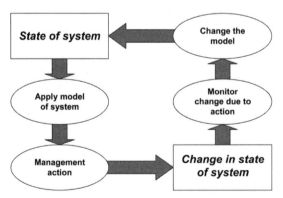

FIGURE 6.2. Feedbacks between policy, management, and research in an adaptive management approach.

- Adaptive management requires flexibility in natural resource management. The uncertainty associated with adaptive management is threatening to the majority of stakeholders (including resource management agencies) who find it difficult to deal with continuous change.
- Resource managers are typically committed to "best management practices" and are reluctant to acknowledge that current practices are not necessarily the best.
- Adaptive management requires a long-term commitment to long-term monitoring and operational research. Such commitments are expensive and hard to sustain over long periods, so management organizations tend to have a poor record on both (Schulte et al. 2006). Outstanding examples such as the U.S. Northwest Forest Plan (Haynes et al. 2006) are rare.

Although few organizations have truly embraced active adaptive management (Stankey et al. 2003; Parr and Andersen 2006), the approach nevertheless represents a logical and potentially powerful way to manage forest and other natural resources. The approach has particular currency in parts of western North America where it is a fundamental component of forest management in some coastal landscapes where logging has been especially controversial (Bunnell et al. 2003; Haynes et al. 2006). Research and monitoring have also been linked through active adaptive management in Kruger National Park in South Africa where "thresholds of potential concern" are used to gauge responses to specific management interventions (Parr and Andersen 2006).

In the particular case of salvage logging, adaptive management would be an excellent approach for continuously improving policy and prescriptions using the best available knowledge (Schmiegelow et al. 2006). Conversely, without attempts to embrace active adaptive management, policies for post-disturbance logging will continue to be controversial and lack an adequate scientific underpinning (Reeves et al. 2006).

In summary, researchers always call for additional research to resolve controversies. However, there really *are* major knowledge gaps on the ecological impacts of salvage logging and the best ways to mitigate such impacts. It is urgent to close these gaps for the following reasons:

- Extensive salvage operations will continue to be proposed following major natural disturbances in North America, Australia, Asia, and Europe (British Columbia Ministry of Forests and Range 2006; Schmiegelow et al. 2006; Victoria Department of Sustainability and Environment 2007).
- Wood salvaged from naturally disturbed areas is likely to continue to be an increasing proportion of harvest volume in some regions (e.g., northwestern North America [McIver and Starr 2000; British Columbia Ministry of Forests and Range 2006] and Canadian provinces such as Quebec [Nappi et al. 2004; fig. 6.3]).
- Climate change is likely to increase the frequency of major disturbance events, such as wildfires (e.g., Franklin et al. 1991; Goetz et al. 2005; Westerling et al. 2006), windstorms and hurricanes (Goldenberg et al. 2001; Stanturf et al. 2007), and insect outbreaks (e.g., Shore et al. 2003). This, in turn, will result in increased demands for salvage logging (Spittlehouse and Stewart 2003).

Of course, there are many research needs beyond the ones listed briefly in the preceding sections. As an example, cross-disciplinary studies melding ecological and economic work will be useful to weigh the socioeconomic benefits generated by salvage logging against the ecological costs of such operations, including the potential to alter timber markets both spatially and temporally (e.g., Holmes 1991; Prestemon and Holmes 2000, 2004). The role of economic analyses in tackling these kinds of is-

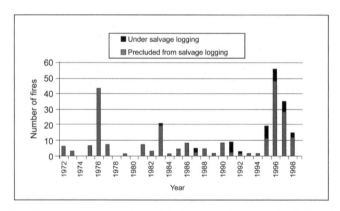

FIGURE 6.3. Increasing proportion of timber supply from salvage logging operations in Quebec, Canada (redrawn from Nappi et al. 2004).

sues was demonstrated in a report by the U.S. Government Accountability Office (2006) that indicated that, while ~$US11 million was spent in salvaging timber from the 2002 Biscuit Burn in Oregon, only ~$US9 million was received from selling the timber. In particular, the cost of road construction (and, in some cases, deconstruction) may be substantial and needs to be included in economic assessments of the returns and costs of salvage logging. Conversely, other studies indicate that where salvage logging is delayed, marked reductions in financial returns can result (Prestemon et al. 2006). There are also concerns about the continuity of timber supplies (and associated economic activity and employment opportunities) when rates of logging are temporarily accelerated because of salvage logging operations (Patriquin et al. 2005).

Recognition of the Role and Importance of Natural Disturbance Regimes

The following need broader recognition by conservation biologists, resource managers, policymakers, and the general public:

- The essential role of natural disturbances in the maintenance of biodiversity and ecosystem processes (Hutto 1995, 2006; Angelstam 1996; Reice 2001).
- The ability of ecosystems to recover, unassisted, from natural disturbances (Turner et al. 2003; Society for Conservation Biology Scientific Panel on Fire in Western U.S. Forests 2005; fig. 6.4).
- The value of recently disturbed areas as uncommon but often critical habitats for particular elements of the biota (Murphy and Lehnhausen 1998; Morissette et al. 2002; Saint-Germain et al. 2004).
- The fact that even though the natural regeneration of vegetation following natural disturbance is sometimes slow (Shatford et al. 2007), the extended period prior to canopy closure may have positive ecological consequences. Human intervention to speed "recovery" may be detrimental to some ecological values and disrupt natural recovery.

There is also a need for greater recognition that the effects of logging in post–naturally disturbed environments can be different from those of natural disturbances in isolation, and different from the logging of

FIGURE 6.4. Rapid recovery after a high-intensity wildfire in shrubland vegetation in December 2003 at Booderee National Park, Jervis Bay Territory, southeastern Australia. A. Two days post-fire. B. After two years (Photos by David Lindenmayer and Christopher MacGregor).

otherwise undisturbed forest (Foster and Orwig 2006; Lindenmayer and Noss 2006).

Although the task of highlighting the ecological importance of disturbance and the value of disturbed and naturally recovering ecosystems is critically important, it will not be easy since most people view disturbances as destructive phenomena. Related to this are human perspectives such as (1) "leaving alone" rather than "cleaning up" runs counter to the psyche of command-and-control efficiency that pervades much of Western society (Hull 2006), and (2) "messy" habitats are considered un-aesthetic relative to those perceived to be "orderly," "undisturbed," and "in balance with the environment" (Noss and Lindenmayer 2006).

The Need for New Language

Terminology is one of the problems associated with the lack of appreciation of the ecological impacts of salvage logging. Dictionary definitions of *salvage* associate it with "recover or save" or "saving of anything from loss or danger" (e.g., Macquarie Dictionary 1989). Although salvage logging removes wood from burned, insect-attacked, or wind-affected areas and can generate economic returns, such practices generally do not help "save" ecological values or "recover" ecosystems or communities—indeed, they often have the opposite effect (Foster and Orwig 2006; Thompson et al. 2007).

Because the term *salvage* is inappropriate, even misleading, from ecological and conservation perspectives, we suggest that policymakers, ecologists, and forest managers need to find an alternative term, such as *post-disturbance logging*, that better describes the activity and its effects on ecosystem processes and on forest biota. Better terms may exist, and we encourage broader dialogue to identify them. There may even be indigenous terms that could be utilized, especially since the psyche of "cleaning up" is a largely Western construct that may not have strong parallels with indigenous cultures.

The terms *catastrophe* and *catastrophic disturbance* are also problematic, particularly when used to describe large-scale natural disturbances (Reice 2001). These are used not only by resource managers, policymakers, and the general public but even by some ecologists (e.g., Ne'eman et al. 1997; Trabaud 2003 cf. Miyanishi 2003). Even the term *stand-replacing disturbance* (e.g., Oliver and Larson 1996) implies the creation of a "clean slate" for ecosystem recovery, which is rarely the case. Such disturbances are often not catastrophic for ecosystem function or biodiversity and may not require "restoration" or "rehabilitation" (i.e., human intervention; see Simon et al. 1994; Burton 2006).

Finally, the word *recover* can sometimes misrepresent what is taking place in naturally disturbed ecosystems. Ecosystem conditions soon after major disturbance can represent unique and extremely important environments in their own right, and indeed populations of some species either only occur in those environments (Hutto 2006) or remain totally unaltered relative to pre-disturbance conditions (Whelan 1995).

Summary

Management objectives for forests are the starting point in determining what kind of post-disturbance activities, including salvage logging, should be undertaken. Ecological, economic, and social objectives all need to be considered. Natural disturbances should not result in automatic or de facto changes in management objectives, as sometimes has been the case.

Salvage logging and other post-disturbance practices can have profound negative impacts on ecological processes and biodiversity. Salvage logging will rarely, if ever, contribute in a direct or positive way to ecological recovery; generally it can be viewed as a tax on ecological recovery that can be large or small depending on how it is conducted.

Ecologically informed policies regarding salvage logging should be developed prior to major natural disturbances so that ad hoc and crisis-mode decision making can be avoided when disturbances do occur. Policies for salvage logging need to include the identification of areas where salvage logging should not occur and modifications to practices where it does occur.

There are large gaps in knowledge about the impacts of salvage logging and the ways to mitigate those impacts. Well-conceived study designs are needed to underpin true experiments, natural experiments, and observational studies to help better understand salvage logging impacts. Wherever possible, such studies should be conducted so as to allow salvage logged areas to be rigorously compared with both undisturbed parts of landscapes and disturbed areas exempt from salvage logging. Policies and management actions associated with salvage logging need to be informed scientifically. An adaptive management approach for formally interconnecting research, policymaking, and on-the-ground management is frequently discussed but only rarely implemented. Nevertheless, it is a logical framework for improving policies and practices for salvage logging in many (if not all) jurisdictions around the world. Adaptive management experiments must be designed and implemented in close partnership with resource managers. This is a fundamental tenet of adaptive management experiments; otherwise, they will "fail the test of management relevance." Implementing true adaptive management will require stakeholders to commit to a culture of continuous improvement in natural resource management and hence acclimatize to an environment of uncertainty. True adaptive management also demands a greater commitment to innovation, monitoring, and flexibility in modifying policies and practices than currently seems apparent in natural resource management.

Concluding Remarks

Climate change is almost certain to result in more frequent, intense, widespread, and novel disturbances. Hence, salvage or post-disturbance logging is likely to become an increasingly common form of timber harvesting in many parts of the world. The practice is also increasingly being encouraged by some government policies. The ecological impacts of salvage logging have the potential to substantially exceed those of green logging, even traditional high-intensity silvicultural systems such as clearcutting followed by even-aged stand management. The ecological impacts of salvage logging, both short term and long term, warrant greater attention than they have received in the past. We hope that this book will contribute to more discussions about this increasingly important topic.

adaptive management   A structured program of improving policies, practices, or standards, whereby the effectiveness of management decisions
(or, preferably, alternative decisions) is monitored and used to improve
future decisions

age class   Group of organisms (e.g., trees, or stands of tree) of more or less
the same age; usually denotes more arbitrary divisions or classes than
cohort

agent (of disturbance)   Type of event that initiates tree mortality, such as
wildfire, windstorm, or insects

aggregated retention   Trees (live or dead) kept standing in clumps or clusters
after logging

back-burning   Using prescribed fire for fuel reduction during the course of
wildland firefighting, with the prescribed fire advancing against the
wind

biodiversity   The diversity of taxa and biological processes found at all levels
of the ecological hierarchy (genes, species, communities, ecosystems)

biological legacy   Life-form, propagule, organic structure or material (or its
footprint) remaining after an ecological disturbance

blackout burning   Using prescribed fire for fuel reduction during the course
of wildland firefighting

boreal forest   Northern, subarctic biome dominated by conifers and successional broadleaf trees, characterized by a prolonged period of subfreezing temperatures

buffer strip   Narrow, continuous patch of green vegetation or unlogged trees, typically prescribed to ameliorate the impact of industrial activities (such as logging) on aquatic ecosystems or visual aesthetics

bycatch   Nontargeted organisms (such as fish or trees) taken during the course of regular harvest or salvage harvest operations

cable system   Forest harvesting equipment and procedures that use steel ropes to transport logs from the stump to the landing, sometimes able to elevate the log partially or wholly off the ground

carbon sequestration   The absorption of carbon dioxide from the air and its incorporation into cellulose, calcium carbonate, or other nongaseous reservoirs

catchment   A basin of water collection; a watershed

cavity tree   A tree, live or dead, containing one or more cavities or hollows

clearcut   (1) To remove all the trees from an area of land; (2) the area of land from which all trees have recently been cut or removed

cohort   Group of organisms of more or less the same age; usually denotes more biologically meaningful divisions than arbitrary **age classes**

conflagration   Violent fire

connectivity   Degree of linkage, particularly between suitable habitats

contingency planning   Preparing protocols and procedures to still meet program goals in the face of various eventualities

crisis management   Planning, policy change, and operations implemented on a short timeline or with a sense of urgency, typically in response to unforeseen circumstances

cumulative impacts   Negative environmental effects resulting from multiple disturbances or multiple industries, often unforeseen from the additive effects of individual disturbances or industries

cyclone   Large storm system, rotating in a counterclockwise direction (viewed from above) and characterized by sustained high-velocity winds

debris flow   Rapid mass movement of rock, soil, and woody material downhill in a slurry of mud and water

dendrochronology   The use of tree rings to date past events

dispersed retention   Trees (live or dead) kept standing as more or less uniformly spaced individuals after logging

disturbance   An event (more or less discrete in time) that causes mortality of the dominant vegetation, thereby altering habitat and releasing resources

disturbance regime   The combination of agents, frequency, and severity (or other attributes such as seasonality or extent) characteristic of disturbance events in an area

disturbance return interval   The period of time between disturbances at a given spot; see **fire cycle** and **fire return interval**

duff   Organic constituents of the forest floor, particularly the lichen, moss, litter, fermentation, and humus layers above any mineral soil

ecosystem process   An ecological dynamic, such as productivity, succession, or decomposition

edge   Boundary between two compositionally or structurally different ecosystems, land uses, or seral stages

endemic   (1) Evolved in the area of residence; (2) low, chronic, or background population levels

epicormic (sprouting or branches)   On the stem or bole of a tree

epidemic   Explosive population growth by some organism, typically a pest or pathogen

equilibrium theory   The "balance of nature" worldview; the premise that species distributions, community composition, and ecosystem processes are finely attuned to their environment in a self-regulating manner

even-aged management   Forestry practice in which more or less uniform cohorts of trees are established and harvested

even-aged stand   An area of forest dominated by a single cohort of trees

fine fuels   Flammable natural materials of small diameter (high surface area), capable of being ignited

firebrand   Live cinder or ember blown from a forest fire

firebreak   A strip or zone in which less flammable materials prevail, or from which fuels have been removed

fire cycle   Fire return interval, or sometimes used more specifically to denote fire return interval inferred from the current age class structure of a forest

fireline   A firebreak constructed during the course of wildland firefighting to contain a fire, consisting of a narrow strip of land from which woody materials, fine fuels, and duff have been removed

fire return interval   The inverse of annual area burned (expressed as a proportion of a given area of land) by wildfires, denoting the average time between fires

fire suppression   The prevention and extinguishment of wildland fires

gap   An opening in the canopy of a forest

hardwood   Broadleaf tree, or the wood from such a tree

harvesting   The systematic collection of a resource, typically one managed for a particular product

hazard-reduction burning   Use of prescribed fire to consume fuels (typically around the perimeter of homes or communities), so that any subsequent wildfire will be stopped or can more easily be controlled

hydrological regime   The characteristic levels and timing of streamflows in a catchment

hurricane   Tropical cyclone initiated in the Atlantic Ocean or Caribbean Sea

hyporheic   Pertaining to the zone of saturated soils near a stream

ice storm   A weather event in which freezing rain adheres to, weighs down, and frequently breaks tree branches

industrial forest   Land base owned or managed by a forest products company for the production of timber or wood fiber

intermediate disturbance hypothesis   The observation that biotic diversity seems to be greatest where disturbance frequencies and intensities are not too low and not too high

invasive   Having a tendency to expand and competitively displace other organisms

iron pan   A dense layer of soil, impervious to water, with clay particles bound together by iron oxides

landing   Area where logs transported from the stump are transferred to trucks, characterized by temporary log piles and repeated traffic (and hence soil compaction) by heavy machinery

landscape (1) Scenic vista; (2) a relatively large area of land consisting of a mosaic of interacting ecosystems, but sharing a relatively uniform climate and disturbance regime

landslide A mass movement characterized primarily by the sudden collapse or downslope movement of relatively dry rock and soil

legacy A structure or impact persisting beyond the life of its originator

life history The characteristic biological processes, requirements, and behaviors of an organism that allow it to survive and reproduce successfully

managed (stand, forest) Pertaining to an area in which human intervention plays a role in disturbing, regenerating, and guiding the development of the ecosystem in question

mass movement Substantial translocation of geological or soil materials, such as a landslide or debris flow

matrix In landscape ecology, the dominant or most connected land cover in which other cover-type patches are embedded; assumed to be multiple-use or commodity-emphasis lands in the context of conservation biology

mechanical disturbance Physical damage to soil or to vegetation in the form of abrasion, displacement, or breakage

mesoscale Of medium or moderate scope, intermediate between microscale and macroscale

mitigation Reduction or offsetting of impact

multi-aged stand A forest stand consisting of two or more cohorts of trees

multiple use The policy of conserving or managing several different resources or values on the same area of land

natural disturbance Forces of nature (such as fires or storms, more or less discrete in time) that cause mortality of the dominant vegetation

nonequilibrial Not tending to return to some previous state; not necessarily in balance with all influential factors

nucleation (1) Starting at and then spreading from a few discrete centers; (2) the onset of a phase transition in a localized region

overland flow Surface runoff

overstory   That portion of a forest stand consisting of dominant mature trees; see **understory**

paludification   Soil development and successional process by which an area becomes boggy, retaining water in peat moss or other vegetation

paradigm   Mental model, approach, and underlying set of assumptions by which a discipline is practiced or a problem addressed

patch   A discrete area of more or less homogeneous properties; often refers to a polygon of vegetation surrounded by some other kind of vegetation or land cover

perturbation   Minor disturbance or disruption, often causing limited mortality, such that conditions return to their previous state in a relatively short period of time

pest   Organism (animal, plant, fungus, microbe) that is detrimental to the growth or survival of some desired crop or ecosystem attribute

phloem   Vascular tissue that transports carbohydrates in plants, known as "inner bark" in trees

plantation   A forest stand created by the planting of tree seedlings or cuttings

preemptive logging   Cutting of living trees in the anticipation that they will soon be killed by some pathogen or agent of disturbance

prescribed burn   Fire set by humans to achieve specific management objectives such as fuel reduction, seedbed preparation, or ecosystem restoration

prescription   Detailed plan for site-specific management actions

primary productivity   The rate at which green plants fix carbon; net primary productivity is expressed as biomass growth; see **productivity**

proactive management   Policies, plans, and actions designed to mitigate the impact of future disturbances and other risks; see **reactive management**

productivity   (1) The rate at which organisms fix carbon or grow, expressed in terms of dry biomass per unit area; (2) the amount of product that can be generated per unit investment of raw material, labor, or capital

protected area   Park or reserve designed to fulfill some conservation role

reactive management   Policies, plans, and actions that are quickly mobilized or changed in response to unforeseen events or circumstances; see **proactive management**

recruitment (1) Increase in population from outside sources; (2) transition of organisms (e.g., tree seedlings) from one size class to a larger one

reforestation Establishment of trees, usually by artificial means (e.g., sowing, planting), after a disturbance that has killed or removed most of the overstory

refugia Locations where organisms are able to survive a disturbance event; see **remnant patch**

rehabilitation Conversion of land cover or vegetation from one type to a more desirable type

regrowth stand A second-growth forest stand; typically homogeneous, fast-growing forest that has established after a relatively recent disturbance

remnant patch Vegetation left largely undisturbed by a disturbance event that affected the area around it; see **refugia**

residual structure The physical characteristics of a forest stand after a disturbance

resilience Ability to recover or maintain functional (compositional or structural) attributes when challenged with a disturbance or perturbation

restoration The managed repair or assisted recovery of an ecosystem after degradation

riparian Associated with streamside or riverside habitats, or those ecological features found at the interface of terrestrial and wetland ecosystems in general

rotation Period of time (planned or actual) between equivalent events, especially managed disturbances such as timber harvesting

salvage logging Removal of dead or damaged tree boles to recover economic value

sanitation logging Removal of trees to control the spread of a pest or pathogen

secondary succession Shift in dominance of species over time in ecosystems that have retained biological legacies after a disturbance (e.g., after a forest fire or cultivation)

sedimentation Deposition of silt resulting from erosional runoff

sclerophyll forest   Wooded terrain dominated by trees and shrubs with relatively small, hard (lignin-rich), often evergreen leaves, growing under a relatively hot dry climate

secondary forest   Regrowth forest; closed tree cover that has developed after relatively recent disturbances, whether natural or human-caused

seed bank   Viable plant seeds, spores, and other propagules residing on a site (whether in the soil, on the forest floor, or in trees)

seedbed   Substrate on which plant seeds are germinating

seed tree   (1) Mature tree that is the source of seeds for forest regeneration; (2) silvicultural system in which a few mature trees are temporarily retained as a source of seeds to establish a new cohort of trees in the area under prescription

selection   (1) Natural or artificial filtering of organisms so that only certain genotypes reproduce; (2) silvicultural system in which individual trees or groups of trees are harvested as part of the systematic management of an uneven-aged stand

seral   Successional

severity   Degree of impact, typically expressed in terms of the amount of local damage to organisms and biological legacies

shade-tolerant   Able to persist in the understory of a forest; able to survive at low light intensities

shelterwood   Silvicultural system, or the overstory layer retained as part of such a system, in which a layer of mature trees is temporarily retained as a source of seeds and protection (from frost or solar radiation, etc.) to establish a new cohort of trees in the area under prescription

silvicultural system   A program of forest stand management, including procedures for harvesting, regeneration, and stand tending

silviculture   Management of forest stands, typically with an emphasis on promoting tree regeneration and growth

site preparation   Procedures to generate seedbeds or microsites favorable for forest regeneration; may include the use of fire, chemical herbicides, or machines

skidding   Transport of logs from the stump to a landing, in which part of the log (e.g., one end) or the whole log is dragged along the ground

skid trail   Snigging track; route used for repeated skidding of logs

slash   (1) To cut shrubs or (undesired) young trees; (2) detached nonmerchantable branches, treetops, and small trees left on-site as a result of forest harvesting and stand-tending activities

snag   Standing dead tree

snigging track   Route used for repeated skidding of logs; see **skid trail**

softwood   Conifer tree, or the wood from the bole of such a tree

stakeholder   Person or organization with an interest in a particular management or policy issue

stand   Area of forest dominated by a relatively uniform species composition and forest age class or stage of forest development

stand-replacing (disturbance)   Forest mortality event that kills most of the overstory trees, such that a new cohort of trees soon dominates

strategy   General approach to achieving a goal

structure   (1) Physical architecture of a forest stand, both vertical and horizontal; (2) relative abundance of different age classes, size classes, or cover types in a forest stand or a broader landscape

stub   Tall stump; lower portion of a tree bole retained intact (typically for biodiversity values) while the upper portion is removed (for its wood value or for safety)

substrate   Material on which organisms are growing

succession   Shift in dominance of species on a site over time

suppression   (1) Reduced growth of an organism as a result of competition; (2) the prevention and extinguishment of wildfires

sustained yield   An approach to the management of renewable resources such that harvest levels do not exceed rates at which the resource grows and renews itself

temperate forest   Closed tree cover found in a moderate climate, characterized by pronounced warm and cool seasons

thicket   Dense cluster of shrubs or young trees

thinning   (1) Natural mortality of suppressed individuals during the course of population (stand) development or at times of stress; (2) managed removal of some proportion of a stand or population

tip-up mound   Elevated microsite of soil and organic matter resulting when the collapse of a tree tears its root mass and associated soil material from the ground

tropical forest   Closed tree cover located between the Tropic of Cancer and the Tropic of Capricorn, or where no subfreezing temperatures are experienced

underburning   Surface fire prescribed to burn the ground cover or understory but leave the overstory alive

underplanting   Transplanting of nursery-grown tree seedlings into stands that have some overstory tree cover

understory   Ground cover, herbaceous vegetation, shrubs, regenerating trees (seedlings, saplings, stump sprouts) and suppressed trees that are found beneath the main forest canopy

uneven-aged management   A silvicultural strategy designed to harvest a sustained yield of logs while recruiting replacement trees and maintaining tree cover within an individual stand; see **selection (2)**

uneven-aged stand   A forest stand consisting of an intimate mixture of trees of all ages, or at least three cohorts

ungulate   Relatively large, herbivorous placental mammals, mostly hoofed (e.g., deer)

value   Importance or priority to some people

variable retention   Exclusion of some mature live trees from a harvesting plan, providing long-term benefits to biodiversity throughout the rotation of the second-growth stand

vegetation   Plant cover, and the relative abundance or dominance of plant species

water bar   Shallow ditch located perpendicularly across a road, designed to intercept runoff from the road surface and divert it off the road

watershed   (1) Catchment area for runoff and subsurface flow supplying water to a given stream or river; (2) the height of land constituting a boundary between adjacent catchment areas

wildfire   Uncontrolled burning of (usually natural or semi-natural) vegetation, including unintentional fires in grassland, savannah, woodland, and forest

wildland   Area of land dominated by natural vegetation, such as native forest, woodland, grassland, tundra, or desert

windstorm   Air movements with sufficient speed or turbulence to cause damage

windthrow   Tree mortality or collapse caused by air movement, or sometimes specifically the uprooting of trees by wind as opposed to stem breakage caused by wind

REFERENCES

Aapala, K., R. Heikkila, and T. Lindholm. 1996. Protecting the diversity of Finnish mires. In *Peatlands in Finland*, H. Vasander, 45–57. Finnish Peatlands Society, Helsinki, Finland.

Achtemeier, G. L. 2001. Simulating nocturnal smoke movement. *Fire Management Today* 61:28–33.

Adamowicz, W. L., and P. J. Burton. 2003. Sustainability and sustainable forest management. In *Towards Sustainable Management of the Boreal Forest*, P. J. Burton, C. Messier, D. W. Smith, and W. L. Adamowicz, 41–64. NRC Research Press, Ottawa, Canada.

Adams, S. B. 2006. Katrina: Boon or bust for freshwater fish communities? *Watershed* Fall–Winter:19–21, 23.

Adams, M. H., and P. M. Attiwill. 1984. The role of *Acacia* spp. in nutrient balance and cycling in regenerating *Eucalyptus regnans* F. Muell. forests. 1. Temporal changes in biomass and nutrient content. *Australian Journal of Botany* 32:205–15.

Agee, J. K. 1993. *Fire Ecology of the Pacific Northwest Forests*. Island Press, Washington, D.C.

Akay, A. E., J. Sessions, P. Bettinger, R. Toupin, and A. Eklund. 2006. Evaluating the salvage value of fire-killed timber by helicopter—Effects of yarding distance and time since fire. *Western Journal of Applied Forestry* 21:102–7.

Alexander, J. D., N. E. Seavy, and P. E. Hosten. 2007. Using conservation plans and bird monitoring to evaluate the ecological effects of management: An example with fuels reduction activities in southwest Oregon. *Forest Ecology and Management* 238:375–83.

Amaranthus, M. P., and D. A. Perry. 1994. The functioning of ectomycorrzhial fungi in the field: Linkages in space and time. *Plant and Soil* 159:133–40.

Amatya, D. M., M. Miwa, C. A. Harrison, C. G. Trettin, and G. Sun. 2006. Hydrology and water quality of two first order forested watersheds in coastal South Carolina. Paper No. 062182 presented at the American Society of Agricultural and Biological Engineers Annual International Meeting, 9–12 July, 2006, Portland, Oregon.

183

American Lands Alliance. 2003. *Restoration or Exploitation? Post-Fire Salvage Logging in America's National Forests*. American Lands Alliance, Washington, D.C., November 2003.

Amiro, B. D., J. M. Chen, and J. Liu. 2000. Net primary productivity following forest fire for Canadian ecoregions. *Canadian Journal of Forest Research* 30:939–47.

Amman, G. D., and K. C. Ryan. 1991. *Insect Infestation of Fire-Injured Trees in the Greater Yellowstone Area*. Research note INT-398. USDA, Forest Service, Intermountain Research Station, Ogden, Utah.

Andrew, N., L. Rodgerson, and A. York. 2000. Frequent fuel-reduction burning: The role of logs and associated leaf litter in the conservation of ant biodiversity. *Austral Ecology* 25:99–107.

Angelstam, P. 1996. The ghost of forest past—Natural disturbance regimes as a basis for reconstruction for biologically diverse forests in Europe. In *Conservation of Faunal Diversity in Forested Landscapes*, R. M. DeGraaf and R. I. Miller, 287–337. Chapman and Hall, London.

Angst, C., and R. Volz. 2002. A decision-support tool for managing storm damaged forests. *Forest and Snow Landscape Research* 77:217–24.

Aronson, R. B., and W. F. Precht. 1995. Landscape pattern of coral reef diversity: A test of the intermediate disturbance hypothesis. *Journal of Experimental and Marine Biology and Ecology* 192:1–14.

Ashton, D. H. 1975. The root and shoot development of *Eucalyptus regnans* F. Muell. *Australian Journal of Botany* 23:867–87.

Ashton, D. H. 1981. Fire in tall open forests. In *Fire and the Australian Biota*, A. M. Gill, J. J. Mott, R. A. Groves, and I. R. Noble, 339–66. Australian Academy of Science, Canberra, Australia.

Ashton, D. H. 1986. Ecology of bryophytic communities in mature *Eucalyptus regnans* F. Muell. forest at Wallaby Creek, Victoria. *Australian Journal of Botany* 34:107–29.

Attiwill, P. M. 1994a. Ecological disturbance and the conservative management of eucalypt forests in Australia. *Forest Ecology and Management* 63:301–46.

Attiwill, P. M. 1994b. The disturbance of forest ecosystems: The ecological basis for conservative management. *Forest Ecology and Management* 63:247–300.

Australian Bureau of Statistics. 2004. *Year Book Australia 1301.0: Environment—Bushfires*. Australian Bureau of Statistics, Canberra, Australia. Available on-line at http://www.abs .gov.au/ausstats/abs@.nsf/0/ccb3f2e90ba779d3ca256dea00053977?OpenDocument [accessed 5 Feb. 2007].

Backer, D. M., S. E. Jensen, and G. R. McPherson. 2004. Impacts of fire suppression activities on natural communities. *Conservation Biology* 18:937–44.

Baird, B. N. 2006. Comment on: post-wildfire logging hinders regeneration and increases fire risk. *Science* 313:615b.

Baker, W. 1992. The landscape ecology of large disturbances in the design and management of nature reserves. *Landscape Ecology* 7:181–94.

Ball, I., D. B. Lindenmayer, and H. P. Possingham. 1999. HOLSIM: a model for simulating hollow availability in managed forest stands. *Forest Ecology and Management* 123:79–194.

Banner, A., P. LePage, J. R. Moran, and A. de Groot. 2005. *The HyP³ Project: Pattern, Process and Productivity in Hyper-maritime Forests of Coastal British Columbia—A Syn-*

thesis of 7-year Results. Special Report Series 10. Research Branch, B. C. Ministry of Forests and Range, Victoria, Canada. Available on-line at http://www.for.gov.bc.ca/hfd/pubs/Docs/Srs/Srs10.htm [accessed 27 Sept., 2007].

Barlow, J., C. A. Peres, L. M. Henriques, P. C. Stouffer, and J. M. Wunderle. 2006. The responses of understorey birds to forest fragmentation, logging and wildfires: An Amazonian synthesis. *Biological Conservation* 128:182–92.

Bartlett, T., M. Butz, and P. Kanowski. 2005. Engaging the community in reforestation after the 2003 Canberra bushfire. Burning issues: The future of forestry. *Proceedings of the Institute of Foresters of Australia National Conference*, Mt. Gambier, 10–14 April 2005. Institute of Foresters of Australia, Mt. Gambier, Australia.

Bascompte, J., and M. A. Rodriguez. 2000. Self-disturbance as a source of spatiotemporal heterogeneity: The case of the tallgrass prairie. *Journal of Theoretical Biology* 204:153–64.

Bayley, P. B. 1995. Understanding large river-floodplain ecosystems. *BioScience* 45:153–58.

Bazzaz, F. A. 1983. Characteristics of populations in relation to disturbance in natural and man-modified ecosystems. In *Disturbance and Ecosystems: Components of Response*, H. A. Mooney and M. Godron, 259–75. Springer-Verlag, New York.

Beatty, S. W., and B. S. Owen. 2005. Incorporating disturbance into forest restoration. In *Restoration of Boreal and Temperate Forests*, J. A. Stanturf and P. Madsen, 61–76. CRC Press, New York.

Beckage, B., and J. I. Stout. 2000. Effects of repeated burning on species richness in a Florida pine savanna: A test of the intermediate disturbance hypothesis. *Journal of Vegetation Science* 11:113–22.

Bejer, B. 1988. The nun moth in European spruce forests. In *Dynamics of Forest Insect Populations: Patterns, Causes, Implications*, A. A. Berryman, 211–31. Plenum Press, New York.

Berg, A., B. Ehnstrom, L. Gustaffson, T. Hallingback, M. Jonsell, and J. Weslien. 1994. Threatened plant, animal and fungus species in Swedish forests: Distribution and habitat associations. *Conservation Biology* 8:718–31.

Bergeron, Y., R. Bradshaw, and O. Engelmark. 1993. Disturbance dynamics in boreal forest: Introduction. *Journal of Vegetation Science* 4:729–32.

Bergeron, Y., B. Harvey, A. Leduc, and S. Gauthier. 1999. Forest management guidelines based on natural disturbance dynamics: Stand- and forest-level considerations. *Forestry Chronicle* 75:49–54.

Bergeron, Y., A. Leduc, B. D. Harvey, and S. Gauthier. 2002. Natural fire regime: A guide for sustainable management of the Canadian boreal forest. *Silva Fennica* 36:81–95.

Bergeron, Y., P. J. H. Richard, C. Carcaillet, S. Gauthier, M. Flannigan, and Y. T. Prairie. 1998. Variability in fire frequency and forest composition in Canada's southeastern boreal forest: A challenge for sustainable forest management. *Conservation Ecology* 2(2):6. Available on-line at http://www.ecologyandsociety.org/vol2/iss2/art6/

Beschta, R. 1978. Long-term patterns of sediment production following road construction and logging in the Oregon Coast range. *Water Resources Research* 14:1011–16.

Beschta, R. 1979. Debris removal and its effects on sedimentation in an Oregon Coast Range stream. *Northwest Science* 53:71–77.

Beschta, R., C. Frissell, R. Gresswell, R. Hauer, J. R. Karr, G. W. Minshall, D. A. Perry, and J. J. Rhodes. 1995. *Wildfire and Salvage Logging. Recommendations for*

Ecologically Sound Post-fire Salvage Management and Other Post-fire Treatments on Federal Lands in the West. Oregon State University Report. Corvallis, Oregon.

Beschta, R., J. J. Rhodes, J. B. Kauffman, R. E. Gresswell, G. W. Minshall, J. R. Karr, D. A. Perry, F. R. Hauer, and C. A. Frissell. 2004. Postfire management on forested public lands of the Western United States. *Conservation Biology* 18:957–67.

Beyers, J. L. 2004. Postfire seeding for erosion control: Effectiveness and impacts on native plant communities. *Conservation Biology* 18:947–56.

Blake, J. G. 1983. Influence of fire and logging on nonbreeding bird communities of ponderosa pine forests. *Journal of Wildlife Management* 46:404–15.

Boose, E. P., M. I. Serrano, and D. F. Foster. 2004. Landscape and regional impacts of hurricanes in Puerto Rico. *Ecological Monographs* 74:335–52.

Borchers, J. G., and D. A. Perry. 1990. Influence of harvesting and residues on fuels and fire management. In *Natural and Prescribed Fire in the Pacific Northwest Forests*, J. D. Walstad, S. R. Radosevich, and D. V. Sandberg, 143–57. Oregon State University Press, Corvallis, Oregon.

Bormann, B. T., H. Spaltenstein, M. H. McClellan, F. C. Ugolini, K. Cromack, and S. M. Nay. 1995. Rapid soil development after windthrow disturbance in pristine forests. *Journal of Ecology* 83:747–57.

Boubée, J., I. Jowlett, S. Nichols, and E. Williams. 1999. *Fish Passage at Culverts: A Review, with Possible Solutions for New Zealand Indigenous Species.* Department of Conservation, Wellington, New Zealand.

Bowman, D. M. 2003. Bushfires: A Darwinian perspective. In *Australia Burning: The Future of Fire in Australia*, G. Cary, D. B. Lindenmayer, and S. Dovers, 3–14. CSIRO Publishing, Melbourne, Australia.

Bradbury, S. M. 2006. Response of the post-fire bryophyte community to salvage logging in boreal mixedwood forests of northeastern Alberta, Canada. *Forest Ecology and Management* 234:313–22.

Bradstock, R. A., J. E. Williams, and A. M. Gill, editors. 2002. *Flammable Australia: The Fire Regimes and Biodiversity of a Continent.* Cambridge University Press, Melbourne, Australia.

Brais, S., P. David, and R. Ouimet. 2000. Impacts of wild fire severity and salvage harvesting on the nutrient balance of jack pine and black spruce boreal stands. *Forest Ecology and Management* 137:231–43.

Brawn, J. D., S. K. Robinson, and F. R. Thompson. 2001. The role of disturbance in the ecology and conservation of birds. *Annual Reviews of Ecology and Systematics* 32:251–76.

Bray, J. R. 1956. Gap phase replacement in a maple-basswood forest. *Ecology* 37:598–600.

Breshears, D. D., N. S. Cobb, P. M. Rich, K. P. Price, C. D. Allen, R. G. Balice, W. H. Romme, J. H. Kastens, M. L. Floyd, J. Belnap, J. J. Anderson, O. B. Meyers, and C. W. Meyer. 2005. Regional vegetation die-off in response to global-change-type drought. *Proceedings of the National Academy of Sciences* 102:15144–48.

British Columbia Ministry of Forests and Range. 2006. *Mountain Pine Beetles in British Columbia.* Available on-line at www.for.gov.bc.ca/hfp/mountain_pine_beetle/

British Columbia Ministry of Forests and Range. 2007. *Forests for Tomorrow: Responding to Catastrophic Wildfires and the Mountain Pine Beetle Epidemic.* Victoria, B. C. Available on-line at http://www.for.gov.bc.ca/hfp/fft/ [accessed 5 February 2007].

Brokaw, N. V., and R. A. Lent. 1999. Vertical structure. In *Managing Biodiversity in Forest Ecosystems*, M. Hunter III, 373–99. Cambridge University Press, Cambridge.

Broncano, M. J., J. Retana, and A. Rodrigo. 2005. Predicting the recovery of *Pinus halepensis* and *Quercus ilex* forests after a large wildfire in north-eastern Spain. *Plant Ecology* 180:47–56.

Brooks, R. T. 2001. Effects of removal of overstorey hemlock from hemlock-dominated forests on eastern redback salamanders. *Forest Ecology and Management* 149:197–204.

Brooks, R. T. 2004. Early regeneration following presalvage cutting of hemlock from hemlock-dominated stands. *Northern Journal of Applied Forestry* 12:12–18.

Brown, K. A., and J. Gurevitch. 2004. Long-term impacts of logging on forest diversity in Madagascar. *Proceedings of the National Academy of Sciences* 101:6045–49.

Brown, R. T., J. K. Agee, and J. F. Franklin. 2004. Forest restoration and fire: Principles in the context of place. *Conservation Biology* 18:903–12.

Bryant, E. A. 2001. *Tsunami: The Underrated Hazard.* Cambridge University Press, Cambridge.

Bryant, E. A. 2003. Tsunami in Australia. *Nature Australia* 27:4–51.

Buddle, C. M., D. W. Langor, G. R. Pohl, and J. R. Spence. 2006. Arthropod responses to harvesting and wildfire: Implications for emulation of natural disturbance in forest management. *Biological Conservation* 128:346–57.

Buddle, C. M., J. R. Spence, and D. W. Langor. 2000. Succession of forest spider assemblages following wildfire and harvesting. *Ecography* 23:424–36.

Bull, E. L., K. B. Aubry, and B. C. Wales. 2001. Effects of disturbance on forest carnivores of conservation concern in eastern Oregon and Washington. *Northwest Science* 75 (Special Issue S1):180–84.

Bull, E. L., and A. D. Partridge. 1986. Methods of killing trees for use by cavity nesters. *Wildlife Society Bulletin* 14:142–46.

Bunnell, F. 1995. Forest-dwelling fauna and natural fire regimes in British Columbia: Patterns and implications for conservation. *Conservation Biology* 9:636–44.

Bunnell, F. 1999. What habitat is an island? In *Forest Wildlife and Fragmentation: Management Implications*, J. Rochelle, L. A. Lehmann, and J. Wisniewski, 1–31. Brill, Leiden, Germany.

Bunnell, F., G. Dunsworth, D. Huggard, and L. Kremsater. 2003. *Learning to Sustain Biological Diversity on Weyerhaeuser's Coastal Tenure.* Weyerhaeuser Company, Vancouver, Canada.

Bunnell, F., and L. Kremsater. 1990. Sustaining wildlife in managed forests. *Northwest Environmental Journal* 6:243–69.

Bunnell, F. L., K. A. Squires, and I. Houde. 2004. *Evaluating Effects of Large-Scale Salvage Logging for Mountain Pine Beetle on Terrestrial and Aquatic Vertebrates.* Mountain Pine Beetle Initiative Working Paper 2004-2. Pacific Forestry Centre, Canadian Forest Service, Victoria, Canada. Available on-line at http://mpb.cfs.nrcan.gc.ca/research/publications_e.html [accessed 8 February 2007].

Burley, S., S. L. Robinson, and J. T. Lundholm. 2008. Post-hurricane vegetation recovery in an urban forest. *Landscape and Urban Planning* 85:111–22.

Burrows, N., and G. Wardell-Johnson. 2003. Fire and plant associations in forested ecosystems of south-west Western Australia. In *Fire in Ecosystems of South-west Western*

Australia: Impacts and Management, I. Abbott and N. Burrows, 225–68. Blackhuys Publishers, Ledien, The Netherlands.

Burton, P. J. 1995. The Mendelian compromise: A vision for equitable land use allocation. *Land Use Policy* 12: 63–68.

Burton, P. J. 2005. Ecosystem management and conservation biology. In *Forestry Handbook for British Columbia,* S. B. Watts and L. Tolland, 307–22. Faculty of Forestry, University of British Columbia, Vancouver, Canada.

Burton, P. J. 2006. Restoration of forests attacked by mountain pine beetle: Misnomer, misdirected, or must-do? *British Columbia Journal of Ecosystems and Management* 7(2):1–10.

Burton, P. J., W. L. Adamowicz, G. F. Weetman, C. Messier, E. Prepas, and R. Tittler. 2003a. The state of boreal forestry and the drive for change. In *Towards Sustainable Management of the Boreal Forest,* P. J. Burton, C. Messier, D. W. Smith, and W. L. Adamowicz, 1–40. NRC Research Press, Ottawa, Canada.

Burton, P. J., D. D. Kneeshaw, and K. D. Coates. 1999. Managing forest harvesting to maintain old growth in boreal and sub-boreal forests. *Forestry Chronicle* 75:623–31.

Burton, P. J., C. Messier, W. Adamowicz, and T. Kuuluvainon. 2006. Sustainable management of Canada's boreal forests: Progress and prospects. *EcoScience* 13:234–48.

Burton, P. J., C. Messier, D. W. Smith, and W. L. Adamowicz, editors. 2003b. *Towards Sustainable Management of the Boreal Forest.* NRC Research Press, Ottawa, Canada.

Burton, P. J., M.-A. Parisien, J. A. Hicke, R. J. Hall, and J. T. Freeburn. 2008. Large fires as agents of ecological diversity in the North American boreal forest. *International Journal of Wildland Fire* (at press).

Byrne, T., C. Stonestreet, and B. Peter. 2006. Characteristics and utilization of post-mountain pine beetle wood in solid wood products. In *The Mountain Pine Beetle: A Synthesis of Biology, Management, and Impacts on Lodgepole Pine,* L. Safranyik and W. R. Wilson, 233–53. Pacific Forestry Centre, Canadian Forest Service, Victoria, Canada.

Calhoun, A. 1999. Forested wetlands. In *Managing Biodiversity in Forest Ecosystems,* M. Hunter III, 300–31. Cambridge University Press, Cambridge.

Campbell, A. J., and M. T. Tanton. 1981. Effects of fire on the invertebrate fauna of soil and litter of a eucalypt forest. In *Fire and the Australian Biota,* A. M. Gill, J. J. Mott, R. A. Groves, and I. R. Noble, 215–41. Australian Academy of Science, Canberra, Australia.

Canadian Council of Forest Ministers. 2007. *Compendium of Canadian Forestry Statistics, National Forestry Database Program.* Canadian Forest Service, Natural Resources Canada, Ottawa, Canada. Available on-line at http://nfdp.ccfm.org/compendium/

Canadian Forest Service. 2006. *State of Canada's Forests, 2005.* Natural Resources Canada, Ottawa, Ontario. Available on-line at http://nrcan.gc.ca/sof/

Canadian Forest Service. 2007. *Carbon Budget Model for the Canadian Forest Sector.* Natural Resources Canada, Ottawa, Canada. Available on-line at http://carbon.cfs.nrcan.gc.ca/downloads_e.html#programs

Cannell, M., and M. Coutts. 1988. Growing in the wind. *New Scientist* 198:42–46.

Carey, A. B., B. R. Lippke, and J. Sessions. 1999. Intentional systems management: Managing forests for biodiversity. *Journal of Sustainable Forestry* 9:83–125.

Carroll, A. L., T. L. Shore, and L. Safranyik. 2006. Direct control: Theory and practice. In *The Mountain Pine Beetle: A Synthesis of Biology, Management, and Impacts on*

Lodgepole Pine, L. Safranyik and W. R. Wilson, 155–72. Pacific Forestry Centre, Canadian Forest Service, Victoria, Canada.

Carroll, A. L., S. W. Taylor, J. Régnière, and L. Safranyik. 2004. Effects of climate change on range expansion by the mountain pine beetle in British Columbia. In *Mountain Pine Beetle Symposium: Challenges and Solutions, October 30–31, 2003, Kelowna, British Columbia, Canada*, Information Report BC-X-399, T. L. Shore, J. E. Brooks, and J. E. Stone, 223–32. Pacific Forestry Centre, Canadian Forest Service, Victoria, Canada.

Carroll, M. S., A. J. Findley, K. A. Blatner, S. Rodrigeuz Menez, S. E. Daniels, and G. B. Walker. 2000. *Social Assessment for the Wenatchee National Forest Wildfires of 1994: Targeted Analysis for the Leavenworth, Entiat, and Chelan Ranger Districts.* USDA Forest Service. Pacific Northwest Research Station. General Technical Report PNW-GTR-479. Portland, Oregon.

Carter, M. C., and C. D. Foster. 2004. Prescribed burning and productivity in southern pine forests: A review. *Forest Ecology and Management* 191:93–109.

Cary, G. 2002. Importance of a changing climate for fire regimes in Australia. In *Flammable Australia: The Fire Regimes and Biodiversity of a Continent*, R. Bradstock, J. Williams, and A. M. Gill, 26–48. Cambridge University Press, Melbourne, Australia.

Cary, G., D. B. Lindenmayer, and S. Dovers, editors. 2003. *Australia Burning: Fire Ecology, Policy and Management Issues.* CSIRO Publishing, Melbourne, Australia.

Chambers, C. L., and J. N. Mast. 2005. Ponderosa pine snag dynamics and cavity excavation following wildfire in northern Arizona. *Forest Ecology and Management* 216:227–40.

Chambers, C. L., W. C. McComb, and J. C. Tappeiner. 1999. Breeding bird responses to three silvicultural treatments in the Oregon Coast Range. *Ecological Applications* 9:171–85.

Chan-McLeod, A. A., and F. Bunnell. 2004. Potential approaches to integrating silvicultural control of mountain pine beetle with wildlife and sustainable management objectives. In *Mountain Pine Beetle Symposium: Challenges and Solutions, October 30–31, 2003, Kelowna, British Columbia, Canada.* Information Report BC-X-399, T. L. Shore, J. E. Brooks, and J. E. Stone, 267–76. Pacific Forestry Centre, Canadian Forest Service, Victoria, Canada.

Chattaway, M. M. 1958. The regenerative powers of certain eucalypts. *Victorian Naturalist* 75:45–46.

Che, S.-H., and K. Woen. 1997. Comparison of plant community structures in cut and uncut areas at burned area of Mt. Gumo-san. *Journal of Korean Forest Society* 86:509–20.

Chesterfield, E. A., J. McCormick, and G. Hepworth. 1991. The effect of low root temperatures on the growth of mountain forest eucalypts in relation to the ecology of *Eucalyptus nitens*. *Proceedings of the Royal Society of Victoria* 103:67–76.

Chou, Y. H., R. A. Minnich, and R. J. Dezzani. 1993. Do fire sizes differ between southern California and Baja California? *Forest Science* 39:835–44.

Christensen, N. L., J. K. Agee, P. F. Brussard, J. Hughes, D. H. Knight, G. W. Minshall, J. M. Peek, S. J. Pyne, F. J. Swanson, J. W. Thomas, S. Wells, S. E. Williams, and H. A. Wright. 1989. Interpreting the Yellowstone fires of 1988. *BioScience* 39:678–85.

Christiansen, E., and A. Bakke. 1988. The spruce bark beetle of Eurasia. In *Dynamics of Forest Insect Populations: Patterns, Causes, Implications*, A. A. Berryman, 479–503. Plenum Press, New York.

Clark, T. W., and S. C. Minta. 1994. *Greater Yellowstone Ecosystem: Prospects for Ecosystem Science, Management and Policy*. Homestead Publishing, Moose, Wyoming.

Cleland, D. T., T. R. Crow, S. C. Saunders, D. I. Dickmann, A. L. Maclean, J. K. Jordan, R. L. Watson, A. M. Sloan, and K. D. Brosofske. 2004. Characterizing historical and modern fire regimes in Michigan (USA): A landscape ecosystem approach. *Landscape Ecology* 19:311–25.

Clements, F. E. 1916. *Plant Succession: An Analysis of the Development of Vegetation*. Publication 242. Carnegie Institute, Washington, D.C.

Coates, K. D., and P. J. Burton. 1997. A gap-based approach for development of silvicultural systems to address ecosystem management objectives. *Forest Ecology and Management* 99:337–54.

Coates, K. D., C. DeLong, P. J. Burton, and D. L. Sachs. 2006. *Abundance of Secondary Structure in Lodgepole Pine Stands Affected by Mountain Pine Beetle. Report for the Chief Forester of British Columbia*. Bulkley Valley Centre for Natural Resources Research and Management, Smithers, Canada.

Cogbill, C. V. 1996. Black growth and fiddlebutts: The nature of old-growth red spruce. In *Eastern Old-growth Forests: Prospects for Rediscovery and Recovery*, M. B. Davies, 113–25. Island Press, Washington, D.C.

Colburn, E. A. 2004. *Vernal Pools: Ecology and Conservation*. McDonald and Woodward Publishing Company, Granville, Ohio.

Collins, S. L., S. M. Glenn, and D. J. Gibson. 1995. Experimental analysis of intermediate disturbance and initial floristic composition: Decoupling cause and effect. *Ecology* 76:486–92.

Commonwealth of Australia and Department of Natural Resources and Environment. 1997. *Comprehensive Regional Assessment—Biodiversity. Central Highlands of Victoria*. The Commonwealth of Australia and Department of Natural Resources and Environment, Canberra, Australia.

Connell, J. H. 1978. Diversity in tropical rainforests and coral reefs. *Science* 199:1302–10.

Connell, J. H., and R. O. Slatyer. 1977. Mechanisms of succession in natural communities and their role in community stability and organization. *American Naturalist* 111:1119–44.

Conner, R. C., T. Adams, B. J. Butler, W. A. Bechold, T. G. Johnson, S. N. Oswalt, G. Smith, S. Will-Wolf, and C. W. Woodall. 2004. *The State of South Carolina's Forests, 2001*. Resource Bulletin SRS-96. USDA Forest Service, Southern Research Station, Asheville, North Carolina.

Conner, R. N., and D. C. Rudolph. 1989. *Red-cockaded Woodpecker Colony Status and Trends on the Angelina, Davy Crockett and Sabine National Forests*. Research Paper SO-250. USDA Forest Service, Southern Forest Experimental Station, New Orleans, Louisiana.

Conner, R. N., D. C. Rudolph, and J. R. Walters. 2001. *The Red-cockaded Woodpecker: Surviving in a Fire-maintained Ecosystem*. University of Texas Press, Austin.

Converse, S. J., G. C. Whiate, K. L. Farris, and S. Zack. 2006. Small mammals and forest fuel reduction: National-scale responses to fire and fire surrogates. *Ecological Applications* 16:1717–29.

Cooper-Ellis, S., D. R. Foster, G. Carlton, and A. Lezberg. 1999. Forest response to catastrophic wind: Results from an experimental hurricane. *Ecology* 80:2683–96.

Covington, W. W. 2003. The evolutionary and historical context. In *Ecological Restoration of Southwestern Ponderosa Pine Forests*, P. Friederici, 26–47. Island Press, Washington, D.C.

Covington, W. W., P. Z. Fulé, M. M. Moore, S. C. Hart, T. E. Kolb, J. N. Mast, S. S. Sackett, and M. R. Wagner. 1997. Restoring ecological health in ponderosa pine forests of the Southwest. *Journal of Forestry* 95(4):23–29.

Coyle, D. R., T. E. Nebeker, E. R. Hart, and W. J. Mattson. 2004. Biology and management of insect pests in North American intensively managed hardwood forest systems. *Annual Review of Entomology* 50:1–29.

Cram, D., T. Baker, and J. Boren. 2006. *Wildland fire effects in silviculturally treated vs. untreated stands of New Mexico and Arizona*. Research Paper RMRS-RP-55. USDA Forest Service, Rocky Mountain Forest and Range Experiment Station, Fort Collins, Colorado.

Cremer, K. W., B. J. Myers, F. van der Duys, and I. Craig. 1977. Silvicultural lessons from the 1974 windthrow in the radiata pine plantations near Canberra. *Australian Forestry* 40:274–92.

Crisafulli, C. M., F. J. Swanson, and V. H. Dale. 2005. Overview of ecological responses to the eruption of Mount St. Helens: 1980–2005. In *Ecological Responses to the Eruption of Mount St. Helens*, V. H. Dale, F. J. Swanson, and C. M. Crisafulli, 287–99. Springer-Verlag, New York.

Cumming, S. G., P. J. Burton, and B. Klinkenberg. 1996. Boreal mixedwood forests may have no "representative" areas: Some implications for reserve design. *Ecography* 19:162–80.

Cumming, S. G., F. K. A. Schmiegelow, and P. J. Burton. 2000. Gap dynamics in boreal aspen stands: Is the forest older than we think? *Ecological Applications* 10:744–59.

Curtis, R. O. 1997. The role of extended rotations. In *Creating a Forestry for the 21st Century*, K. A. Kohm and J. F. Franklin, 165–70. Island Press, Washington, D.C.

Dale, V. H., L. A. Joyce, S. McNulty, R. P. Nielson, M. P. Ayres, M. D. Flannigan, P. J. Hanson, L. C. Irland, A. E. Lugo, C. J. Peterson, D. Simberloff, F. J. Swanson, B. J. Stocks, and B. M. Wotton. 2001. Climate change and forest disturbances. *BioScience* 51:723–34.

Dale, V., F. J. Swanson, and C. M. Crisafulli. 2005. *Ecological Responses to the 1980 Eruptions of Mount St. Helens*. Springer-Verlag, New York.

Davies, E. J. 1980. Useless? The case against contorta. *Scottish Forestry* 34:110–13.

Davis, L. S., K. N. Johnson, P. S. Bettinger, and T. E. Howard. 2001. *Forest Management to Sustain Ecological, Economic, and Social Values*. McGraw-Hill, New York.

DeGraaf, R., and I. Miller, editors. 1996. *Conservation of Faunal Diversity in Forested Landscapes*. Chapman and Hall, London.

DellaSala, D. A., J. R. Karr, T. Schoennagel, D. Perry, R. F. Noss, D. Lindenmayer, R. Beschta, R. L. Hutto, M. E. Swanson, and J. Evans. 2006a. Post-fire logging debate ignores many issues. *Science* 314:51–52.

DellaSala, D. A., G. Nagle, R. Fairbanks, D. Odion, J. E. Williams, J. R. Karr, C. Frissell, and T. Ingalsbee. 2006b. *The Facts and Myths of Post-fire Management: A Case Study of the Biscuit Fire, Southwest Oregon*. Report to World Wildlife Fund, Klamath-Siskiyou Program, Ashland, Oregon.

DellaSala, D. A., S. B. Reid, T. J. Frest, J. R. Strittholt, and D. M. Olsen. 1999. A global perspective on the biodiversity of the Klamath-Siskiyou ecoregion. *Natural Areas Journal* 19:300–19.

del Moral, R., and L. C. Bliss. 1993. Mechanisms of primary succession: Insights resulting from the eruption of Mount St. Helens. *Advances in Ecological Research* 24:1–66.

DeLong, S. C., and W. B. Kessler. 2000. Ecological characteristics of mature forest remnants left by wildfire. *Forest Ecology and Management* 131:93–106.

deMaynadier, P. G., and M. L. Hunter. 1995. The relationship between forest management and amphibian ecology: A review of the North American literature. *Environmental Reviews* 3:230–61.

Denslow, J. S. 1987. Tropical rainforest gaps and tree species diversity. *Annual Review of Ecology and Systematics* 18:431–51.

Diaz, N., and D. Apostol. 1992. *Forest Landscape Analysis and Design: A Process for Developing and Implementing Land Management Objectives for Landscape Patterns.* R6 ECO-TP-043-92. USDA Forest Service, Pacific Northwest Research Station, Portland, Oregon. Available on-line from http://www.srs.fs.fed.us/pubs/viewpub.jsp?index=3048.

Dobson, J. E., R. M. Rush, and R. W. Peplies. 1990. Forest blowdown and lake acidification. *Annals of the Association of American Geographers* 80:343–61.

Donato, D. C., J. B. Fontaine, J. L. Campbell, W. D. Robinson, J. B. Kauffman, and B. E. Law. 2006a. Post-wildfire logging hinders regeneration and increases fire risk. *Science* 311:352.

Donato, D. C., J. B. Fontaine, J. L. Campbell, W. D. Robinson, J. B. Kauffman, and B. E. Law. 2006b. Response to comments on: Post-wildfire logging hinders regeneration and increases fire risk. *Science* 313:615c.

Donovan, G. H. 2004. Consumer willingness to pay a price premium for standing-dead Alaska yellow-cedar. *Forest Products Journal* 54:38–42.

Donovan, G. H., and T. C. Brown. 2007. Careful what you wish for: The legacy of Smokey Bear. *Frontiers in Ecology and Environment* 5:73–79.

Douglas, T., and P. J. Burton. 2004. Integrating ecosystem restoration into forest management in British Columbia, Canada. In *Restoration of Boreal and Temperate Forests*, J. A. Stanturf and P. Madsen, 423–44. CRC Press, Boca Raton, Florida.

Drapeau, P., A. Nappi, J.-F. Giroux, A. Leduc, and J.-P. Savard. 2002. Distribution patterns of birds associated with snags in natural and managed eastern boreal forests. In *Ecology and Management of Dead Wood in Western Forests*, B. Laudenslayer and B. Valentine, 193–205. General Technical Report PSW-GTR-181. USDA Forest Service, Pacific Southwest Research Station, Albany, California.

Drever, C. R., G. Peterson, C. Messier, Y. Bergeron, and M. Flannigan. 2006. Can forest management based on natural disturbances maintain ecological resiliency? *Canadian Journal of Forest Research* 36:2285–99.

Drury, W. H., and I. C. T. Nesbit. 1973. Succession. *Journal of the Arnold Arboretum* 54:331–68.

Dugay, S. M., K. Aril, M. Hooper, and M. J. Lechowicz. 2001. Ice storm damage and early recovery in an old growth forest. *Environmental Monitoring and Assessment* 67:97–108.

Duncan, R. P. 1993. Flood disturbance and the coexistence of species in a lowland podocarp forest, south Westland, New Zealand. *Journal of Ecology* 81:403–16.

Eberhart, K. E., and P. M. Woodard. 1987. Distribution of residual vegetation associated with large fires in Alberta. *Canadian Journal of Forest Research* 117:1207–12.

Eggler, W. A. 1948. Plant communities in the vicinity of the volcano El Parícutin, Mexico, after two and a half years of eruption. *Ecology* 29:415–36.

Elliott, K. J., S. L. Hitchcock, and L. Krueger. 2002. Vegetation response to large scale disturbance in a southern Appalachian forest: Hurricane Opal and salvage logging. *Journal of the Torrey Botanical Society* 129:48–59.

Emanuel, K. 2005. Increasing destructiveness of tropical cyclones over the past 30 years. *Nature* 436:686–88.

Enoksson, B., Angelstam, P., and Larsson, K. 1995. Deciduous forest and resident birds: The problem of fragmentation within a coniferous forest landscape. *Landscape Ecology* 10:267–75.

Esseen, P.-E., B. Ehnström, L. Ericson, and K. Sjöberg. 1997. Boreal forest. *Ecological Bulletins* 46:6–47.

Etheridge, D. A., D. A. MacLean, R. G. Wagner, and J. S. Wilson. 2006. Effects of intensive forest management on stand and landscape characteristics in northern New Brunswick, Canada (1945–2027). *Landscape Ecology* 21:509–24.

Fail, J. 1999. Production and decomposition rates of a coastal plain forest following the impact of Hurricane Hugo. *Journal of the Elisha Mitchell Scientific Society* 115:47–54.

Fenger, M., T. Manning, J. Cooper, S. Guy, and P. Bradford. 2006. *Wildlife and Trees in British Columbia.* Lone Pine Publishing, Edmonton, Canada.

Fenton, N., N. Lecomte, S. Légaré, and Y. Bergeron. 2005. Paludification of black spruce (*Picea mariana*) forests of eastern Canada: Potential factors and management implications. *Forest Ecology and Management* 213:151–59.

Fischer, W. C., and B. R. McClelland. 1983. *A Cavity-Nesting Bird Bibliography Including Related Titles on Forest Snags, Fire, Insects, Diseases and Decay.* General Technical Report INT-140. USDA Forest Service, Intermountain Forest and Range Experiment Station, Ogden, Utah.

Flannigan, M. D., K. A. Logan, B. D. Amiro, W. R. Skinner, and B. J. Stocks. 2005. Future area burned in Canada. *Climatic Change* 72:1–16.

Fleming, R. A., J. N. Candau, and R. S. McAlpine. 2002. Landscape-scale analysis of interactions between insect defoliation and forest fire in central Canada. *Climatic Change* 55:251–72.

Florence, R. 1996. *Ecology and Silviculture of Eucalypts.* CSIRO Publishing, Melbourne, Australia.

Folke, C., S. Carpenter, B. Walker, M. Scheffer, T. Elmqvist, L. Gunderson, and C. S. Holling. 2004. Regime shifts, resilience, and biodiversity in ecosystem management. *Annual Review of Ecology Evolution and Systematics* 35:557–81.

Food and Agriculture Organization. 2001. *Global Forest Fire Assessment, 1990–2000.* Forestry Department, Food and Agriculture Organization of the United Nations, Rome. Available on-line at http://www.fao.org/docrep/006/ad653e/ad653e00.htm [accessed 21 June 2007].

Forest Ecosystem Management Assessment Team. 1993. *Forest Ecosystem Management: An Ecological, Economic, and Social Assessment.* USDA Forest Service, Pacific Northwest Research Station, Portland, Oregon.

Forman, R. T. 1995. *Landscape Mosaics.* Cambridge University Press, Cambridge.

Forman, R. T. T., D. Sperling, J. A. Bissonette, A. P. Clevenger, C. D. Cutshall, V. H. Dale, L. Fahrig, R. France, C. R. Goldman, K. Heanue, J. A. Jones, F. J. Swanson, T. Turrentine, and T. C. Winter, editors. 2002. *Road Ecology: Science and Solutions.* Island Press, Washington, D.C.

Foster, D. R., and J. D. Aber, editors. 2004. *Forests in Time: The Environmental Consequences of 1000 Years of Change in New England.* Yale University Press, New Haven, Connecticut.

Foster, D. R., J. B. Aber, J. M. Melillo, R. D. Bowden, and F. A. Bazzaz. 1997. Forest response to disturbance and anthropogenic stress. *BioScience* 47:437–45.

Foster, D. R., and E. R. Boose. 1992. Patterns of forest damage resulting from catastrophic wind in central New England, U. S. A. *Journal of Ecology* 80:79–98.

Foster, D. R., and E. R. Boose. 1995. Hurricane disturbance regimes in temperate and tropical forest ecosystems. In *Wind Effects of Forests, Trees and Landscapes,* M. Couts, 305–39. Cambridge University Press, Cambridge.

Foster, D. R., and D. A. Orwig. 2006. Preemptive and salvage harvesting of New England forests: When doing nothing is a viable alternative. *Conservation Biology* 20:959–70.

Franklin, J. F. 1990. Biological legacies: A critical management concept from Mount St. Helens. *Transactions of the 55th North American Wildlife and Natural Resource Conference*: 216–19.

Franklin, J. F. 2004. *Comments on Draft Environmental Impact Statement for Biscuit Recovery Project.* College of Forest Resources, University of Washington, Seattle.

Franklin, J. F. 2005. *Testimony before House Subcommittee on Forests and Forest Health's Legislation Hearing on HR-2000. November 10, 2005.* College of Forest Resources, University of Washington, Seattle.

Franklin, J. F. 2006. *Testimony before House Subcommittee on Forests and Forest Health Scientific Research and the Knowledge-base Concerning Forest Management Following Wildfires and Other Major Disturbances. Material presented to Field Hearing in Medford, Oregon, February 24, 2006.* College of Forest Resources, University of Washington, Seattle.

Franklin, J. F., and J. K. Agee. 2003. Forging a science-based national forest fire policy. *Issues in Science and Technology* 20:59–66.

Franklin, J. F., D. E. Berg, D. A. Thornburgh, and J. C. Tappeiner. 1997. Alternative silvicultural approaches to timber harvest: Variable retention harvest systems. In *Creating a Forestry for the 21st Century,* K. A. Kohm and J. F. Franklin, 111–39. Island Press, Washington, D.C.

Franklin, J. F., and R. T. Forman. 1987. Creating landscape patterns by forest cutting: Ecological consequences and principles. *Landscape Ecology* 1:5–18.

Franklin, J. F., P. M. Frenzen, and F. J. Swanson. 1995. Re-creation of ecosystems at Mount St. Helens: Contrasts in artificial and natural processes. In *Rehabilitating Damaged Ecosystems,* Second Edition, J. Cairns, 287–333. Lewis Publishers, Boca Raton, Florida.

Franklin, J. F., D. B. Lindenmayer, J. A. MacMahon, A. McKee, J. Magnusson, D. A. Perry, R. Waide, and D. R. Foster. 2000. Threads of continuity: Ecosystem disturbances, biological legacies and ecosystem recovery. *Conservation Biology in Practice* 1:8–16.

Franklin, J. F., and J. A. MacMahon. 2000. Messages from a mountain. *Science* 288:1183–85.

Franklin, J. F., J. A. MacMahon, F. J. Swanson, and J. R. Sedell. 1985. Ecosystem responses to the eruption of Mount St. Helens. *National Geographic Research* Spring 1985:198–216.

Franklin, J. F., T. A. Spies, R. van Pelt, A. Carey, D. Thornburgh, D. R. Berg, D. B. Lindenmayer, M. Harmon, W. Keeton, and D. C. Shaw. 2002. Disturbances and the structural development of natural forest ecosystems with some implications for silviculture. *Forest Ecology and Management* 155:399–423.

Franklin, J. F., F. J. Swanson, M. E. Harmon, D. A. Perry, T. A. Spies, V. H. Dale, A. McKee, W. K. Ferrell, J. E. Means, S. V. Gregory, J. D. Lattin, T. D. Schowalter, and D. Larsen. 1991. Effects of global climate change on forests in northwestern North America. *Northwest Environmental Journal* 7:233–54.

Franklin, J. F., and R. van Pelt. 2004. Spatial aspects of structural complexity in old growth forests. *Journal of Forestry* 102(3):22–27.

Fraser, E., S. Landhausser, and V. Lieffers. 2004. The effect of fire severity and salvage logging traffic on regeneration and early growth of Aspen suckers in north-central Alberta. *Forestry Chronicle* 80:251–56.

Frelich, L. E. 2005. *Forest Dynamics and Disturbance Regimes: Studies from Temperate Evergreen-Deciduous Forests.* Cambridge University Press, Cambridge.

Frelich, L. E., and C. G. Lorimer. 1991. Natural disturbance regimes in hemlock-hardwood forests of the Upper Great Lakes region. *Ecological Monographs* 61:145–64.

Friederici, P., editor. 2003. *Ecological Restoration of Southwestern Ponderosa Pine Forests.* Island Press, Washington, D.C.

Fries, C., O. Johansson, B. Petterson, and P. Simonsson. 1997. Silvicultural models to maintain and restore natural stand structures in Swedish boreal forests. *Forest Ecology and Management* 94:89–103.

Frothingham, E. H. 1924. Some silvicultural aspects of the chestnut blight situation. *Journal of Forestry* 22:861–72.

Gayer, K. 1886. *Der gemischte wald—seine Begründung and Pflege, insbesondere durch Horst- und Gruppenwirtschaft.* Verlag Paul Parey, Berlin, Germany.

Geertsema, M., and J. J. Pojar. 2007. Influence of landslides on biophysical diversity—A perspective from British Columbia. *Geomorphology* 89:55–69.

Gibbons, P., and D. B. Lindenmayer. 2002. *Tree Hollows and Wildlife Conservation in Australia.* CSIRO Publishing, Melbourne, Australia.

Gill, A. M. 1975. Fire and the Australian flora: A review. *Australian Forestry* 38:4–25.

Gill, A. M. 1999. Biodiversity and bushfires: An Australia-wide perspective on plant-species changes after a fire event. In *Australia's Biodiversity—Responses to Fire*, A. M. Gill, J. C. Z. Woinarski, and A. York, 9–53. Biodiversity Technical Paper No. 1. Environment Australia, Canberra.

Gill, A. M., and M. A. McCarthy. 1998. Intervals between prescribed fires in Australia: What intrinsic variation should apply? *Biological Conservation* 85:161–69.

Gill, A. M., J. C. Z. Woinarski, and A. York. 1999. *Australia's Biodiversity—Responses to Fire.* Biodiversity Technical Paper No. 1. Environment Australia, Canberra.

Goetz, S. J., A. G. Bunn, G. J. Fiske, and R. A. Houghton. 2005. Satellite-observed photo-

synthetic trends across boreal North America associated with climate and fire disturbance. *Proceedings of the National Academy of Sciences* 102:13521–25.

Goldenberg, S. B., C. W. Landsea, A. M. Mestas-Nuñez, and W. M. Gray. 2001. The recent increase in Atlantic hurricane activity: Causes and consequences. *Science*
293:474–79.

Golley, F. B. 1974. Structural and functional properties as they influence ecosystem stability. *Proceedings of the First International Congress of Ecology*. The Hague, Netherlands, Sept. 8–14, 1974, 97–102.

Gooday, P., P. Whish-Wilson, and L. Weston. 1997. Regional Forest Agreements. Central
Highlands of Victoria. In *Australian Forest Products Statistics*. Australian Bureau of
Agricultural and Resource Economics, 1–11. Canberra, Australia. September Quar-
ter, 1997.

Gordon, G., A. S. Brown, and T. Pulsford. 1988. A Koala (*Phascolartos cinereus* Goldfuss)
population crash during drought and heatwave conditions in southwestern Queensland. *Australian Journal of Ecology* 13:451–61.

Government of the ACT. 2006. *Lower Cotter Catchment: Draft Strategic Management
Plan*. ACT Government, Canberra, Australia.

Government of Victoria. 1986. *Government Statement No. 9. Timber Industry Strategy*.
Government Printer, Melbourne, Australia.

Government of Victoria. 2003. *Report of the Inquiry into the 2002–2003 Victorian Bushfires*. State Government of Victoria, Melbourne, Australia.

Greenberg, C. H. 2002. Response of white-footed mice (*Peromyscus leucopus*) to coarse
woody debris and microsite use in southern Appalachian tree falls. *Forest Ecology and
Management* 164:57–66.

Greenberg, C. H., L. D. Harris, and D. G. Neary. 1995a. A comparison of bird communities in burned and salvaged-logged clearcut, and forested Florida Sand Pine Scrub.
Wilson Bulletin 107:40–54.

Greenberg, C. H., and A. McGrane. 1996. A comparison of relative abundance and biomass of ground-dwelling arthropods under different forest management practices. *Forest Ecology and Management* 89:31–41.

Greenberg, C. H., D. G. Neary, and L. D. Harris. 1994. Effects of high-intensity wildfire
and silvicultural treatments on reptile communities in Sand Pine Scrub. *Conservation
Biology* 8:1047–57.

Greenberg, C. H., D. G. Neary, L. D. Harris, and S. P. Linda. 1995b. Vegetation recovery
following high-intensity wildfire and silvicultural treatments in Sand Pine Scrub.
American Midland Naturalist 133:149–63.

Greenberg, C. H., and M. C. Thomas. 1995. Effects of forest management practices on
terrestrial coleopteran assemblages in Sand Pine Scrub. *Florida Entomologist* 78:271–
85.

Greene, D. F., S. Gauthier, J. Noel, M. Rousseau, and Y. Bergeron. 2006. A field experiment to determine the effect of post-fire salvage on seedbeds and tree regeneration.
Frontiers in Ecology and Environment 4:69–74.

Gregory, S. V. 1997. Riparian management in the 21st century. In *Creating a Forestry for
the 21st Century*, K. A. Kohm and J. F. Franklin, 69–85. Island Press, Washington,
D.C.

Gresham, C. A. 2004. Loblolly pine saplings affected by Hurricane Hugo retained growth.
In *Proceedings of the 12th Biennial Southern Research Converence*, K. F. Connor,

General Technical Report SRS-71, 196–98. USDA Forest Service, Southern Research Station, Asheville, North Carolina.

Griesbauer, H., and S. Green. 2006. Examining the utility of advance regeneration for reforestation and timber production in unsalvaged stands killed by the mountain pine beetle: Controlling factors and management implications. *British Columbia Journal of Ecosystems and Management* 7:81–92.

Groffman, P. M., J. S. Baron, T. Blett, A. J. Gold, I. Goodman, L. H. Gunderson, B. M. Levinson, M. A. Palmer, H. W. Paerl, G. D Peterson, N. L. Poff, D. W. Regeski, J. F. Reynolds, M. G. Turner, K. C. Weathers, and J. Wiens. 2006. Ecological thresholds: The key to successful environmental management or an important concept with no practical application? *Ecosystems* 9:1–13.

Groom, M. J., G. K. Meffe, and C. R. Carroll. 2005. *Principles of Conservation Biology*. Sinauer Associates, Sunderland, Massachusetts.

Grove, S. J., and J. L. Hanula, editors. 2006. *Insect Biodiversity and Dead Wood: Proceedings of a Symposium for the 22nd International Congress of Entomology*. General Technical Report SRS-93. USDA Forest Service, Southern Research Station, Asheville, North Carolina.

Grove, S. J., J. F. Meggs, and A. Goodwin. 2002. *A Review of Biodiversity Conservation Issues Relating to Coarse Woody Debris Management in the Wet Eucalypt Production Forests of Tasmania*. Forestry Tasmania, Hobart, Australia.

Haeussler, S., and Y. Bergeron. 2004. Range of variability in boreal aspen plant communities after wildfire and clear-cutting. *Canadian Journal of Forest Research* 34:274–88.

Haeussler, S., and D. Kneeshaw. 2003. Comparing forest management to natural processes. In *Towards Sustainable Management of the Boreal Forest*, P. J. Burton, C. Messier, D. W. Smith, and W. L. Adamowicz, 307–68. NRC Research Press, Ottawa, Canada.

Haggard, M., and W. L. Gaines. 2001. Effects of stand replacement fire and salvage logging on a cavity-nesting bird community in eastern Cascades, Washington. *Northwest Science* 75:387–96.

Haila, Y., I. K. Hanski, J. Niemelä, P. Puntilla, S. Raivio and H. Tukia. 1993. Forestry and the boreal fauna: Matching management with natural forest dynamics. *Annales Zoologici Fennici* 31:187–202.

Haim, A., and I. Izhaki. 1994. Changes in rodent community during recovery from fire: Relevance to conservation. *Biodiversity and Conservation* 3:573–85.

Halpern, C. B. 1988. Early successional pathways and the resistance and resilience of forest communities. *Ecology* 69:1703–15.

Halpern, C. B., and T. A. Spies. 1995. Plant species diversity in natural and managed forests of the Pacific Northwest. *Ecological Applications* 5:913–34.

Hansen, A., and J. Rotella. 1999. Abiotic factors. In *Managing Biodiversity in Forest Ecosystems*, M. Hunter III, 161–209. Cambridge University Press, Cambridge.

Hansen, A. J., T. A. Spies, F. J. Swanson, and J. L. Ohmann. 1991. Conserving biodiversity in managed forests. *BioScience* 41:382–92.

Hansen, S. B. 1983. *The Effects of the Baxter Park Fire on the Vegetation and Soils of Several Coniferous Stands*. MSc. Thesis, University of Maine, Orono, Maine.

Hanson, C. T., and M. P. North. 2006. Post-fire epicormic branching in Sierra Nevada *Abies concolor* (white fir). *International Journal of Wildland Fire* 15:31–35.

Hanson, J. J., and J. D. Stuart. 2005. Vegetation responses to natural and salvage logged edges in Douglas-fir/hardwood forests. *Forest Ecology and Management* 214:266–78.

Harcourt, C. 1992. Tropical moist forests. In *Global Diversity: Status of the Earth's Living Resources*. World Conservation Monitoring Centre, 256–75. Chapman and Hall, London.

Harmon, M. E., J. F. Franklin, F. J. Swanson, P. Sollins, S. V. Gregory, J. D. Lattin, N. H. Anderson, S. P. Cline, N. G. Aumen, J. R. Sedell, G. W. Lienkaemper, K. J. Cromack, and K. W. Cummins. 1986. Ecology of coarse woody debris in temperate ecosystems. *Advances in Ecological Research* 15:133–302.

Harper, K. A., S. E. Macdonald, P. J. Burton, J. Chen, K. D. Brosofske, S. C. Saunders, E. S. Euskirchen, D. Roberts, M. S. Jaiteh, and P.-A. Esseen. 2005. Edge influence on forest structure and composition in fragmented landscapes. *Conservation Biology* 19:768–82.

Harrington, G. N., and K. D. Sanderson. 1994. Recent contraction of wet sclerophyll forest in the wet tropics of Queensland due to invasion by rainforest. *Pacific Conservation Biology* 1:3319–27.

Haskins, K. E., and C. A. Gehring. 2004. Long-term effects of burning slash on plant communities and arbuscular mycorrhizae in a semi-arid woodland. *Journal of Applied Ecology* 41:379–88.

Hawkes, B. C., S. W. Taylor, C. Stockdale, T. L. Shore, S. J. Beukema, and D. Robinson. 2005. Predicting mountain pine beetle impacts on lodgepole pine stands and woody debris characteristics in a mixed severity fire regime using PrognosisBC and the fire and fuels extension. In *Mixed Severity Fire Regimes: Ecology and Management, November 17–19, 2004, Spokane, Washington*, Vol. AFE MISC03, L. Lagene, J. Zelnik, S. Cadwallader, and B. Hughes, 123–35. Washington State University Coop Extension Service / The Association for Fire Ecology, Pullman, Washington.

Haynes, R. W., B. T. Bormann, D. C. Lee, and J. R. Martin. 2006. *Northwest Forest Plan—The First 10 Years (1993–2003): Synthesis of Monitoring and Research Results.* General Technical Report PNW-GTR-651. USDA Forest Service, Pacific Northwest Research Station, Portland, Oregon.

Heliövaara, K., and R. Väisänen. 1984. Effects of modern forestry on northwestern European forest invertebrates: A synthesis. *Acta Forestalia Fennica* 189:1–32.

Helms, J. A. 1998. *The Dictionary of Forestry.* Society of American Foresters, Bethesda, Maryland.

Helvey, J. D. 1980. Effects of a north central Washington wildfire on runoff and sediment production. *Water Resources Bulletin* 16:627–34.

Helvey, J. D., A. R. Tiedemann, and T. D. Anderson. 1985. Plant nutrient losses by soil erosion and mass movement after wildfire. *Journal of Soil and Water Conservation* 40:168–73.

Higgs, P., and B. J. Fox. 1993. Interspecific competition: A mechanism for rodent succession after fire in wet heathland. *Australian Journal of Ecology* 18:193–201.

Hilmo, O., H. Holien, and H. Hytteborn. 2005. Logging strategy influences colonization of common chlorolichens on branches of *Picea abies*. *Ecological Applications* 15:983–86.

Hobbs, R. J., S. Arico, J. Aronson, J. S. Baron, P. Bridgewater, V. Cramer, P. R. Epstein, J. J. Ewel, C. A. Klink, A. E. Lugo, D. Norton, D. Ojima, D. M. Richardson, E. W.

Sanderson, F. Valladres, M. Vila, R. Zamora, and M. Zobel. 2006. Novel ecosystems: Theoretical and management aspects of the new ecological world order. *Global Ecology and Biogeography* 15:1–7.

Hobson, K. A., and J. Schieck. 1999. Changes in bird communities in boreal mixedwood forest: Harvest and wildfire effects over 30 years. *Ecological Applications* 9:849–63.

Holbrook, S. H. 1943. *Burning an Empire.* Macmillan Company, New York.

Holling, C. S., editor. 1978. *Adaptive Environmental Assessment and Management.* International Series on Applied Systems Analysis 3, International Institute for Applied Systems Analysis. John Wiley and Sons, Toronto, Canada.

Holling, C. S. 1992. The role of forest insects in structuring the boreal landscape. In *A Systems Analysis of the Global Boreal Forest,* H. H. Shugart, R. Leemans, and G. B. Bonan, 170–95. Cambridge University Press, Cambridge.

Holling, C. S., and M. Meffe. 1996. Command and control and the pathology of natural resource management. *Conservation Biology* 10:328–37.

Holling, C. S., D. W. Schindler, B. W. Walker, and J. Roughgarden. 1995. Biodiversity in the functioning of ecosystems: An ecological primer and synthesis. In *Biodiversity Loss: Economic and Ecological Issues,* D. Pimental, 44–83. Cambridge University Press, Cambridge.

Holmes, T. P. 1991. Price and welfare effects of catastrophic forest damage from southern pine beetle epidemics. *Forest Science* 37:500–16.

Holtam, B. W., editor. 1971. *Windblow of Scottish Forests in January 1968.* Forestry Commission Bulletin No. 45. Her Majesty's Stationary Office, Edinburgh.

Hooper, R. G., W. E. Taylor, and S. C. Loeb. 2004. Long-term efficacy of artificial cavities for red-cockaded woodpeckers: Lessons learned from Hurricane Hugo. In *Red-Cockaded Woodpecker: Road to Recovery,* R. Costa and D. J. Daniels, 430–38. Hancock House Publishers, Blaine, Washington.

Horton, S. P., and R. W. Mannan. 1988. Effects of prescribed fire on snags and cavity-nesting birds in southeastern Arizona. *Wildlife Society Bulletin* 16:37–44.

How, R. A., J. L. Barnett, A. J. Bradley, W. F. Humphreys, and R. Martin. 1984. The population biology of *Pseudocheirus peregrinus* in a *Leptospermum laevigatum* thicket. In *Possums and Gliders,* A. P. Smith and I. D. Hume, 261–68. Surrey Beatty and Sons, Sydney, Australia.

Hoyt, J. S., and S. J. Hannon. 2002. Habitat associations of black-backed and three-toed woodpeckers in the boreal forest of Alberta. *Canadian Journal of Forest Research* 32:1881–88.

Huggett, R., and J. Cheesman. 2002. *Topography and the Environment.* Prentice-Hall, London.

Hughes, J., and R. Drever. 2001. *Salvaging Solutions. Science-based Management of British Columbia's Pine Beetle Outbreak.* David Suzuki Foundation, Forest Watch of British Columbia Society, and Canadian Parks and Wilderness Society. Vancouver, Canada.

Hull, R. B. 2006. *Infinite Nature.* University of Chicago Press, Chicago.

Hunter, M. L. 1993. Natural fire regimes as spatial models for managing boreal forests. *Biological Conservation* 65:115–20.

Hunter, M. L., editor. 1999. *Managing Biodiversity in Forest Ecosystems.* Cambridge University Press, Cambridge.

Hunter, M. L. 2007. Core principles for using natural disturbance regimes to inform land-scape management. In *Managing and Designing Landscapes for Conservation: Moving from Perspectives to Principles*, D. B. Lindenmayer and R. J. Hobbs, 408–22. Blackwell Publishing, Oxford.

Hutto, R. 1995. Composition of bird communities following stand-replacement fires in northern Rocky Mountain conifer forests. *Conservation Biology* 9:1041–58.

Hutto, R. L. 2006. Toward meaningful snag-management guidelines for postfire salvage logging in North American conifer forests. *Conservation Biology* 20:984–93.

Hutto, R. L., and S. M. Gallo. 2006. The effects of post-fire salvage logging on birds. *The Condor* 108:817–31.

Hyvärinen, E., J. Kouki, and P. Martikainen. 2006. Fire and green-tree retention in conservation of red-listed and rare deadwood-dependent beetles in Finnish boreal forests. *Conservation Biology* 20:1711–19.

Imbeau, L., M. Mönkkönen, and A. Descrochers. 2001. Long-term effects of forestry on birds of the eastern Canadian boreal forests: A comparison with Fennoscandia. *Conservation Biology* 15:1151–62.

Inbar, M., L. Wittenberg, and M. Tamir. 1997. Soil erosion and forest management after wildfire in a Mediterranean woodland, Mt. Carmel, Israel. *International Journal of Wildland Fire* 7:285–94.

Inions, G., M. T. Tanton, and S. M. Davey. 1989. Effects of fire on the availability of hollows in trees used by the common brushtail possum, *Trichosurus vulpecula* Kerr 1792, and the ringtail possum, *Pseudocheirus peregrinus* Boddaerts 1785. *Australian Wildlife Research* 16:449–58.

Irland, L. C. 1998. Ice storm 1998 and the forests of the Northeast. *Journal of Forestry* 96(9):32–40.

Isaac, L. A., and G. S. Meagher. 1938. *Natural Reproduction on the Tillamook Burn Four Years after Fire*. USDA Forest Service Pacific Northwest Research Station, Portland, Oregon.

Jackson, R. B., E. G. Jobbagy, R. Avissar, S. B. Roy, D. J. Barrett, C. W. Cook, K. A. Farley, D. C. le Maitre, B. A. McCarl, and B. C. Murray. 2005. Trading water for carbon with biological sequestration. *Science* 310:1944–47.

Jäggi, C., and B. Baur. 1999. Overgrowing forest as a possible cause for the local extinction of *Vipera aspis* in the northern Swiss Jura mountains. *Amphibia-Reptilia* 20:25–34.

James, F. C., C. A. Hess, and D. Kufrin. 1997. Species-centered environmental analysis: Indirect effects of fire history on red-cockaded woodpeckers. *Ecological Applications* 7:118–29.

James, F. C., P. M. Richards, C. A. Hess, K. E. McCluney, E. L. Walters, and M. S. Schrader. 2003. Sustainable forestry for the red-cockaded woodpecker ecosystem. In *Red-Cockaded Woodpecker: Road to Recovery*, R. Costa and S. J. Daniels, 60–69. Hancock House, Blaine, Washington.

James, I. L., and D. A. Norton. 2002. Helicopter-based natural forest management for New Zealand's rimu (*Dacrydium cupressinum*, Podocarpaceae) forests. *Forest Ecology and Management* 155:337–46.

James, S. E., and R. T. M'Closky. 2003. Lizard microhabitat and fire fuel management. *Biological Conservation* 114:293–97.

Jõgiste, K., W. K. Moser, and M. Mandre. 2005. Disturbance dynamics and ecosystem-based forest management. *Scandinavian Journal of Forest Research* 20(suppl.):2–4.

Johnson, D. W., J. F. Murphy, R. B. Susfalk, T. G. Caldwell, W. W. Miller, R. F. Walker, and R. F. Powers. 2005. The effects of wildfire, salvage logging, and post-fire N fixation on the nutrient budgets of a Sierran forest. *Forest Ecology and Management* 220:155–65.

Johnson, E. A. 1992. *Fire and Vegetation Dynamics: Studies from the North American Boreal Forest.* Cambridge University Press, Cambridge.

Johnson, R. H., R. R. Sharitz, P. M. Dixon, D. S. Segal, and R. L. Schneider. 1994. Woody plant regeneration in four floodplain forests. *Ecological Monographs* 64:45–84.

Kafka, V., S. Gauthier, and Y. Bergeron. 2001. Fire impacts and crowning in the boreal forest: Study of a large wildfire in western Quebec. *International Journal of Wildland Fire* 10:119–27.

Karr, J. R., J. J. Rhodes, G. W. Minshall, F. R. Hauer, R. L. Beschta, C. A. Frissell, and D. A. Perry. 2004. The effects of postfire salvage logging on aquatic ecosystems in the American West. *BioScience* 54:1029–33.

Karraker, N. E., and H. H. Welsh. 2006. Long-term impacts of even-aged timber management on abundance and body condition of terrestrial amphibians in northwestern California. *Biological Conservation* 131:132–40.

Kavanagh, R. P., and R. J. Turner. 1994. Birds in eucalypt plantations: The likely role of retained habitat trees. *Australian Birds* 28:32–41.

Keenan, R. J., and J. P. Kimmins. 1993. The ecological effects of clear-cutting. *Environmental Reviews* 1:121–44.

Keith, D., J. Williams, and J. Woinarski. 2002. Fire management and biodiversity conservation: Key approaches and principles. In *Flammable Australia: The Fire Regimes and Biodiversity of a Continent*, R. Bradstock, J. Williams, and A. M. Gill, 401–25. Cambridge University Press, Cambridge.

Kellas, J. D., and A. J. M. Hateley. 1991. Management of dry sclerophyll forests in Victoria. I. The low elevation mixed forests. In *Forest Management in Australia*, F. H. McKinnell and J. E. D. Fox, 142–62. Surrey Beatty, Chipping Norton, Australia.

Khetmalas, M. B., K. N. Egger, H. B. Massicotte, L. E. Tackaberry, and M. J. Clapperton. 2002. Bacterial diversity associated with subalpine fir (*Abies lasiocarpa*) ectomycorrhizae following wildfire and salvage logging in central British Columbia. *Canadian Journal of Microbiology* 48:611–25.

King, A. P. 1963. *The Influences of Colonization on the Forests and the Prevalence of Bushfires in Australia.* CSIRO Division of Physical Chemistry, Melbourne, Australia.

Kizlinski, M. L., D. A. Orwig, R. C. Cobb, and D. R. Foster. 2002. Direct and indirect ecosystem consequences of an invasive pest on forests dominated by eastern hemlock. *Journal of Biogeography* 29:1489–1503.

Klock, G. O. 1975. Impact of postfire salvage logging systems on soils and vegetation. *Journal of Soil and Water Conservation* 30:78–81.

Kohnle, U., and K. von Teuffel. 2004. Norway spruce: Production of large-sized timber still economically sensible in southwestern Germany. *Allgemeine Forst und Jagzeitung* 175:171–82.

Kotliar, N. B., S. J. Heijl, R. L. Hutto, V. A. Saab, C. P. Melcher, and M. E. McFadzen. 2002. Effects of fire and post-fire salvage logging on avian communities in conifer-dominated forests of the western United States. *Studies in Avian Biology* 25:49–64.

Kotliar, N. B., P. L. Kennedy, and K. Ferree. 2007. Avifaunal responses to fire in southwestern montane forests along a burn severity gradient. *Ecological Applications* 17: 491–507.

Krankina, O. N., and M. E. Harmon. 1995. Dynamics of the dead wood carbon pool in northwestern Russian boreal forests. *Water, Air and Soil Pollution* 82:227–38.

Kulakowski, D., and T. T. Veblen. 2007. Effect of prior disturbance on the extent and severity of wildfire in Colorado subalpine forests. *Ecology* 88:759–69.

Kurz, W. A., and M. J. Apps. 1999. A 70-year retrospective analysis of carbon fluxes in the Canadian forest sector. *Ecological Applications* 9:526–47.

Kurz, W. A., M. Apps, E. Banfield, and G. Stinson. 2002. Forest carbon accounting at the operational scale. *Forestry Chronicle* 78:672–79.

Kuuluvainen, T. 2002. Introduction—Disturbance dynamics in boreal forests: Defining the ecological basis of restoration and management of biodiversity. *Silva Fennica* 36:5–10.

Kuuluvainen, T., H. Tukia, and K. Aapala. 2004. Ecological restoration of forested ecosystems in Finland. In *Restoration of Boreal and Temperate Forests*, J. A. Stanturf and P. Madsen, 285–98. CRC Press, Boca Raton, Florida.

Kyker-Snowman, T. 2000. Managing the shift from water yield to water quality on Boston's water supply watersheds. In *Drinking Water from Forests and Grasslands*, G. E. Dissmeyer, 212–14. General Technical Report SRS-39. USDA Forest Service, Southern Research Station, Washington, D.C.

Lamberson, R. H., B. R. Noon, C. Voss, and R. McKelvey. 1994. Reserve design for territorial species: The effects of patch size and spacing on the viability of the northern spotted owl. *Conservation Biology* 8:185–95.

Land Conservation Council. 1994. *Final Recommendations. Melbourne Area. District 2 Review.* Land Conservation Council, Melbourne, Australia.

Landres, P. B., P. Morgan, and F. J. Swanson. 1999. Evaluating the utility of natural variability concepts in managing ecological systems. *Ecological Applications* 9:1179–88.

Larson, A. J., and J. F. Franklin. 2005. Patterns of conifer tree regeneration following an autumn wildfire event in the western Oregon Cascade Range. *Forest Ecology and Management* 218:25–36.

Lavigne, F., and Y. Gunnell. 2006. Land cover change and abrupt environmental impacts on Javan volcanoes, a long-term perspective on recent events. *Regional Environmental Change* 6:86–100.

Law, B. E., D. Turner, J. Campbell, O. J. Sun, S. Van Tuyl, W. D. Ritts, and W. B. Cohen. 2004. Disturbance and climate effects on carbon stocks and fluxes across western Oregon USA. *Global Change Biology* 10:1429–44.

Leathwick, J. R., and N. D. Mitchell. 1992. Forest pattern, climate and volcanism in central North Island, New Zealand. *Journal of Vegetation Science* 3:603–16.

Lecomte, N., M. Simard, and Y. Bergeron. 2006. Effects of fire severity and initial tree composition on stand structural development in the coniferous boreal forest of northwestern Quebec, Canada. *EcoScience* 13:152–63.

Lee, K. N. 1999. Appraising adaptive management. *Conservation Ecology* 3:3. Available on-line at http://www.ecologyandsociety.org/vol3/iss2/art3/

Lenihan, J. M., R. Drapek, D. Bachelet, and R. P. Neilson. 2003. Climate change effect on vegetation distribution, carbon, and fire in California. *Ecological Applications* 13:1667–81.

Lertzman, K. P. 1992. Patterns of gap-phase replacement in a subalpine, old-growth forest. *Ecology* 73:657–69.

Lewis, K., D. Thompson, I. Hartley, and S. Pasca. 2006. *Wood Decay and Degradation in Standing Lodgepole Pine* (Pinus contorta var. latifolia Engelm.) *Killed by Mountain Pine Beetle* (Dendroctonus ponderosae Hopkins: Coleoptera). Mountain Pine Beetle Initiative Working Paper 2006-11. Pacific Forestry Centre, Canadian Forest Service, Victoria, Canada.

Lieffers, V. J., C. Messier, P. J. Burton, J.-C. Ruel, and B. E. Grover. 2003. Nature-based silviculture for sustaining a variety of boreal forest values. In *Towards Sustainable Management of the Boreal Forest*, P. J. Burton, C. Messier, D. W. Smith, and W. L. Adamowicz, 481–530. NRC Research Press, Ottawa, Canada.

Lindemann, J. D., and W. L. Baker. 2001. Attributes of blowdown patches from a severe wind event in the southern Rocky Mountains, USA. *Landscape Ecology* 16:313–25.

Lindenmayer, D. B. 2006. Salvage harvesting: Past lessons and future issues. *Forestry Chronicle* 82:1–6.

Lindenmayer, D. B., A. W. Claridge, A. M. Gilmore, D. Michael, and B. D. Lindenmayer. 2002. The ecological role of logs in Australian forest and the potential impacts of harvesting intensification on log-using biota. *Pacific Conservation Biology* 8:121–40.

Lindenmayer, D. B., R. B. Cunningham, and C. F. Donnelly. 1997. Tree decline and collapse in Australian forests: Implications for arboreal marsupials. *Ecological Applications* 7:625–41.

Lindenmayer, D. B., R. B. Cunningham, C. F. Donnelly, and J. F. Franklin. 2000. Structural features of Australian old growth montane ash forests. *Forest Ecology and Management* 134:189–204.

Lindenmayer, D. B., R. B. Cunningham, and M. A. McCarthy. 1999a. The conservation of arboreal marsupials in the montane ash forests of the Central Highlands of Victoria, south-eastern Australia. VIII. Landscape analysis of the occurrence of arboreal marsupials in the montane ash forests. *Biological Conservation* 89:83–92.

Lindenmayer, D. B., R. B. Cunningham, and R. Peakall. 2005. On the recovery of populations of small mammals in forest fragments following major population reduction. *Journal of Applied Ecology* 42:649–58.

Lindenmayer, D. B., R. B. Cunningham, M. T. Tanton, and A. P. Smith. 1990. The conservation of arboreal marsupials in the montane ash forests of the Central Highlands of Victoria, south-east Australia. II. The loss of trees with hollows and its implications for the conservation of Leadbeaters possum *Gymnobelideus leadbeateri* McCoy (Marsupialia: Petauridae). *Biological Conservation* 54:133–45.

Lindenmayer, D. B., and J. Fischer. 2006. *Habitat Fragmentation and Landscape Change: An Ecological and Conservation Synthesis*. Island Press, Washington, D.C.

Lindenmayer, D. B., and J. Fischer. 2007. Tackling the habitat fragmentation panchreston. *Trends in Ecology and Evolution* 22(3):127–32.

Lindenmayer, D. B., D. Foster, J. F. Franklin, M. Hunter, R. Noss, F. Schiemegelow, and D. Perry. 2004. Salvage harvesting after natural disturbance. *Science* 303:1303.

Lindenmayer, D. B., and J. F. Franklin. 1997. Forest structure and sustainable temperate forestry: A case study from Australia. *Conservation Biology* 11:1053–68.

Lindenmayer, D. B., and J. F. Franklin. 2002. *Conserving Forest Biodiversity: A Comprehensive Multiscaled Approach*. Island Press, Washington, D.C.

Lindenmayer, D. B., J. F. Franklin, and J. Fischer. 2006. Conserving forest biodiversity: A checklist for forest and wildlife managers. *Biological Conservation* 129:511–18.

Lindenmayer, D. B., J. F. Franklin, and D. Foster. 2004b. Salvage harvesting fire-damaged wet eucalypt forests—some ecological perspectives. *Australian Forestry* 67:131–36.

Lindenmayer, D. B., and R. Hobbs, editors. 2007. *Managing and Designing Landscapes for Conservation: Moving from Perspectives to Principles.* Blackwell Scientific, Oxford.

Lindenmayer, D. B., R. D. Incoll, R. B. Cunningham, and C. F. Donnelly. 1999b. Attributes of logs in the mountain ash forests in south-eastern Australia. *Forest Ecology and Management* 123:195–203.

Lindenmayer, D. B., B. Mackey, I. Mullins, M. A. McCarthy, A. M. Gill, R. B. Cunningham, and C. F. Donnelly. 1999c. Stand structure within forest types—Are there environmental determinants? *Forest Ecology and Management* 123:55–63.

Lindenmayer, D. B., and M. A. McCarthy. 2001. The spatial distribution of non-native plant invaders in a pine-eucalypt mosaic in south-eastern Australia. *Biological Conservation* 102:77–87.

Lindenmayer, D. B., and M. A. McCarthy. 2002. Congruence between natural and human forest disturbance—An Australian perspective. *Forest Ecology and Management* 155:319–35.

Lindenmayer, D. B., T. W. Norton, and M. T. Tanton. 1991. Differences between the effects of wildfire and clearfelling in montane ash forests of Victoria and its implications for fauna dependent on tree hollows. *Australian Forestry* 53:61–68.

Lindenmayer, D. B., and R. Noss. 2006. The impacts of salvage harvesting on biodiversity and ecosystem processes: A brief overview. *Conservation Biology* 20:949–58.

Lindenmayer, D. B., and K. Ough. 2006. Salvage harvesting in the montane ash forests of the Central Highlands of Victoria, south-eastern Australia. *Conservation Biology* 20:1005–15.

Lindenmayer, D. B., and C. R. Tambiah. 2005. Novel perspectives on the Boxing Day Tsunami. *Conservation Biology* 19:991.

Linder, P., and L. Östlund. 1998. Structural changes in three mid-boreal Swedish forest landscapes, 1885–1996. *Biological Conservation* 85:9–19.

Logan, J. A., and J. A. Powell. 2001. Ghost forests, global warming, and the mountain pine beetle (Coleoptera: Scolytidae). *American Entomologist* 47:160–72.

Luke, R. H., and A. G. McArthur. 1977. *Bushfires in Australia.* Australian Government Publishing Service, Canberra, Australia.

Lunney, D., editor. 2004. *Conservation of Australia's Forest Fauna.* Royal Zoological Society of New South Wales, Mosman, Sydney, Australia.

Luppold, W. G., and P. E. Sendak. 2004. Analysis of the interaction between timber markets and the forest resources of Maine. *Northern Journal of Applied Forestry* 21:135–43.

Lynch, H. J., R. A. Renkin, R. L. Crabtree, and P. R. Moorcroft. 2006. The influence of previous mountain pine beetle (*Dendroctonus ponderosae*) activity on the 1988 Yellowstone Fires. *Ecosystems* 9:1318–27.

Lynott, R. E., and O. P. Cramer. 1966. Detailed analysis of the 1962 Columbus Day windstorm in Oregon and Washington. *Monthly Weather Review* 94:105–17.

Macdonald, S. E. 2007. Effects of partial post-fire salvage harvesting on vegetation communities in the boreal mixedwood forest region of northeastern Alberta, Canada. *Forest Ecology and Management* 239:21–31.

Macfarlane, M. A., J. Smith, and K. Lowe. 1998. *Leadbeater's Possum Recovery Plan. 1998–2002*. Department of Natural Resources and Environment. Government of Victoria, Melbourne, Australia.

Mackensen, J., J. Bauhus, and E. Webber. 2003. Decomposition rates of coarse woody debris—A review with particular emphasis on Australian tree species. *Australian Journal of Botany* 51:27–37.

Mackey, S. M., and P. M. Cornish. 1982. Effects of wildfire and logging on the hydrology of small catchments near Eden, NSW. In *Proceedings of 1st National Symposium on Forest Hydrology*, 111–17. Sydney, Australia.

Mackey, B. G., D. B. Lindenmayer, A. M. Gill, M. A. McCarthy, and J. Lindesay. 2002. *Wildlife, Fire and Future Climate: A Forest Ecosystem Analysis*. CSIRO Publishing, Melbourne, Australia.

Mackie, R. S. 2000. *Island Timber: A Social History of the Comox Logging Company*. Sono Nis Press, Winlaw, British Columbia.

Macquarie Dictionary. 1989. *Macquarie Concise Dictionary*. The Macquarie Library Pty Ltd, Sydney, Australia.

Mallik, A. V., and B. A. Roberts. 1994. Natural regeneration of *Pinus resinosa* on burned and unburned sites in Newfoundland. *Journal of Vegetation Science* 5:179–86.

Marcot, B. G. 1997. Biodiversity of old forests of the West: A lesson from our elders. In *Creating a Forestry for the 21st Century*, K. A. Kohm and J. F. Franklin, 87–105. Island Press, Washington, D.C.

Marsinko, A., T. J. Straka, and J. L. Bauman. 1993. Hurricane Hugo—A South Carolina update. *Journal of Forestry* 91(9):9–17.

Martínez-Sánchez, J. J., P. Ferrandis, J. de las Heras, and J. M. Herranz. 1999. Effect of burnt wood removal on the natural regeneration of *Pinus halepensis* after fire in a pine forest in Tus Valley (SE Spain). *Forest Ecology and Management* 123:1–10.

Maser, C. 1996. The loss of ecological reason and moral restraint. *International Journal of Ecoforestry* 12:176–87.

May, R. M. 1977. Thresholds and breakpoints in ecosystems with a multiplicity of stable states. *Nature* 269:471–77.

May, S. A. 2001. *Aspects of the Ecology of the Cat, Dog and Fox in the South-east Forests of N. S. W.* Ph. D thesis, The Australian National University, Canberra, Australia.

May, S. A., and T. W. Norton. 1996. Influence of fragmentation and disturbance on the potential impact of feral predators on native fauna in Australian forest ecosystems. *Wildlife Research* 23:387–400.

Mazurek, M. J., and W. J. Zielinski. 2004. Individual legacy trees influence vertebrate wildlife diversity in commercial forests. *Forest Ecology and Management* 193:321–34.

McCarthy, M. A., and M. A. Burgman. 1995. Coping with uncertainty in forest wildlife planning. *Forest Ecology and Management* 74:23–36.

McCarthy, M. A., A. M. Gill, and D. B. Lindenmayer. 1999. Fire regimes in mountain ash forest: Evidence from forest age structure, extinction models and wildlife habitat. *Forest Ecology and Management* 124:193–203.

McCarthy, M. A., and D. B. Lindenmayer. 1998. Development of multi-aged mountain ash (*Eucalyptus regnans*) forest under natural disturbance regimes: Implications for wildlife conservation and logging practices. *Forest Ecology and Management* 104:43–56.

McCarthy, M. A., and D. B. Lindenmayer. 2005. *Risks of Fire and the Management of*

Catchments for Timber Production and Urban Water Supply. CRES Working Paper 2005/01. Centre for Resource & Environmental Studies, The Australian National University, Canberra, Australia.

McCarthy, M. A., and D. B. Lindenmayer. 2006. Info-gap decision theory for accessing the management of catchments for timber production and urban water supply. *Environmental Management* 39:553–62.

McHugh, P. 1991. *State of Resources, Uses and Values. Dandenong (Yarra Forests) Management Area.* Department of Conservation and Environment, Melbourne, Australia.

McIntosh, R. P. 1985. *The Background of Ecology: Concept and Theory.* Cambridge University Press, Cambridge.

McIver, J. D., and R. McNeil. 2006. Soil disturbance and hill-slope sediment transport after logging of a severely burned site in northeastern Oregon. *Western Journal of Applied Forestry* 21:123–33.

McIver, J. D., and R. Ottmar. 2007. Fuel mass and stand structure after post-fire logging of a severely burned ponderosa pine forest in northeastern Oregon. *Forest Ecology and Management* 238:268–79.

McIver, J. D., and L. Starr, technical editors. 2000. *Environmental Effects of Postfire Logging: Literature Review and Annotated Bibliography.* PNW-GTR-486. USDA Forest Service, Pacific Northwest Research Station, Portland, Oregon.

McIver, J. D., and L. Starr. 2001. A literature review on the environmental effects of post-fire logging. *Western Journal of Applied Forestry* 16:159–68.

McNeeley, J. A., L. E. Neville, and M. Rejmanek. 2003. When is eradication a sound investment? *Conservation in Practice* 4:30–41.

McNulty, S. G. 2002. Hurricane impacts on US forest carbon sequestration. *Environmental Pollution* 116:S17–S24.

McRae, D. J., L. C. Duchesne, B. Freedman, T. J. Lynham, and S. Woodley. 2001. Comparisons between wildfire and forest harvesting and their implications in forest management. *Environmental Reviews* 9:223–60.

Metzeling, L., T. Doeg, and W. O'Conner. 1995. The impact of salinisation and sedimentation on aquatic biota. In *Conserving Biodiversity: Threats and Solutions,* R. A. Bradstock, T. A. Auld, D. A. Keith, R. T. Kingsford, D. Lunney, and D. P. Sivertsen, 126–36. Surrey Beatty, Chipping Norton, Australia.

Milledge, D. R., C. L. Palmer, and J. L. Nelson. 1991. Barometers of change: The distribution of large owls and gliders in mountain ash forests of the Victorian Central Highlands and their potential as management indicators. In *Conservation of Australia's Forest Fauna,* D. Lunney, 55–65. Royal Zoological Society of N. S. W., Sydney, Australia.

Millennium Ecosystem Assessment. 2005. *Millennium Ecosystem Assessment Synthesis Report.* World Resources Institute, Washington, D.C. Available on-line at http://www.millenniumassessment.org

Miller, A., and P. Rusnock. 1993. The rise and fall of the silvicultural hypothesis in spruce budworm (*Choristoneura fumiferana*) management in eastern Canada. *Forest Ecology and Management* 61:171–89.

Millward, A. A., and C. E. Kraft. 2004. Physical influences of landscape on a large-extent ecological disturbance: The northeastern North American ice storm of 1998. *Landscape Ecology* 19:99–111.

Minshall, G. W. 2003. Responses of stream benthic macroinvertebrates to fire. *Forest Ecology and Management* 178:155–61.

Mitchell, R. J., M. L. Hunter, and B. J. Palik. 2002. Natural disturbance as a guide to silviculture. *Forest Ecology and Management* 155:315–17.

Mitchell, S. J. 2007. The December 15, 2006 windstorm: What happened and will Stanley Park Recover? *Discovery: Vancouver Natural History Society* 36:15–18.

Miyanishi, K. 2003. Towards a sounder fire ecology. *Frontiers in Ecology and the Environment* 1:275–76.

Mladenoff, D. J., J. White, J. Pastor, and T. R. Crow. 1993. Comparing spatial pattern in unaltered old-growth and disturbed forest landscapes. *Ecological Applications* 3:294–306.

Molino, J., and D. Sabatier. 2001. Tree diversity in tropical rain forests: A validation of the intermediate disturbance hypothesis. *Science* 294:1702–4.

Moore, M. M., D. W. Huffman, P. Z. Fulé, W. W. Covington, and J. E. Crouse. 2004. Comparison of historical and contemporary forest structure and composition on permanent plots in southwestern ponderosa pine forests. *Forest Science* 50:162–76.

Morgan, P., C. C. Hardy, T. W. Swetnam, M. G. Rollins, and D. G. Long. 2001. Mapping fire regimes across time and space: Understanding coarse and fine-scale fire patterns. *International Journal of Wildland Fire* 10:329–42.

Morissette, J. L., T. P. Cobb, R. M. Brigham, and P. C. James. 2002. The response of boreal forest songbird communities to fire and post-fire harvesting. *Canadian Journal of Forest Research* 12:2169–83.

Moritz, M. A., J. E. Keeley, E. A. Johnson, and A. A. Schaffner. 2004. Testing a basic assumption of shrubland management fire: How important is fuel age? *Frontiers in Ecology and Environment* 2:67–72.

Morley, S., C. Grant, R. J. Hobbs, and V. Cramer. 2004. Long-term impact of prescribed burning on the nutrient status of rehabilitated bauxite mines in Western Australia. *Forest Ecology and Management* 190:227–39.

Morrison, M. L., B. G. Marcot, and R. W. Mannan. 2006. *Wildlife–Habitat Relationships. Concepts and Applications.* Island Press, Washington, D.C.

Morrison, P. H., and F. J. Swanson. 1990. *Fire History and Pattern in a Cascade Mountain Landscape.* General Technical Report PNW-GTR-254. USDA Forest Service, Pacific Northwest Research Station, Portland, Oregon.

Morrissey, C. A., C. A. Albert, P. L. Dods, W. R. Cullen, V. W.-M. Lai, and J. E. Elliott. 2007. Arsenic accumulation in bark beetles and forest birds occupying mountain pine beetle infested stands treated with monosodium methanearsonate. *Environmental Science and Technology* 41(4):1494–1500.

Mueck, S. G., K. Ough, and J. C. Banks. 1996. How old are wet forest understories? *Australian Journal of Ecology* 21:345–48.

Murphy, E. C., and W. A. Lehnhausen. 1998. Density and foraging ecology of woodpeckers following a stand replacement fire. *Journal of Wildlife Management* 62:1359–72.

Muzika, R. M., and A. M. Liebold. 2000. A critique of silvicultural approaches to managing defoliating insects in North America. *Agricultural and Forest Entomology* 2:97–105.

Naesset, E. 1999. Decomposition rate constants of *Picea abies* logs in southeastern Norway. *Canadian Journal of Forest Research* 29:372–81.

Naiman, R. J., and R. E. Bilby, editors. 1998. *River Ecology and Management: Lessons from the Pacific Coastal Ecoregion.* Springer-Verlag, New York.

Nappi, A., P. Drapeau, J. F. Giriouux, and J. F. Savard. 2003. Snag use by foraging black-backed woodpeckers (*Picoides articus*) in a recently burned eastern boreal forest. *Auk* 120:505–11.

Nappi, A., P. Drapeau, and J. F. Savard. 2004. Salvage logging after wildfire in the boreal forest: Is it becoming a hot issue for wildlife? *Forestry Chronicle* 80:67–74.

Ne'eman, G., A. Perevolotsky, and G. Schiller. 1997. The management implications of the Mt. Carmel research project. *International Journal of Wildland Fire* 7:343–50.

Newton, M., S. Fitzgerald, R. Rose, P. W. Adams, S. D. Tesch, J. Sessions, T. Atzet, R. Powers, and C. Skinner. 2006. Comment on post-wildfire logging hinders regeneration and increases fire risk. *Science* 313:615a.

Niemelä, J. 2003. Finnish forestry in transition—Finding ways to sustainable forest management. In *Towards Forest Sustainability*, D. B. Lindenmayer and J. F. Franklin, 111–26. CSIRO Publishing, Melbourne, Australia, and Island Press, Washington, D.C.

Nilsson, S. 1974. Changes in the quality of windblown trees left in the stand. *Instituitonen for Skogsteknik* 72:1–52.

Nilsson, S. 1975. Salvage of windthrown forest. *Instituitonen for Skogsteknik* 84:1–96.

Noble, I. R., and R. O. Slatyer. 1980. The use of vital attributes to predict successional changes in plant communities subject to recurrent disturbances. *Vegetatio* 43:5–21.

Noble, W. S. 1977. *Ordeal by Fire: The Week a State Burned Up.* Hawthorn Press, Melbourne, Australia.

Norton, D. A. 1989. Tree windthrow and forest soil turnover. *Canadian Journal of Forest Research* 19:386–89.

Noss, R. F., P. Beier, W. Covington, E. Grumbine, D. B. Lindenmayer, J. Prather, F. Schmiegelow, T. Sisk, and D. Vosick. 2006. Integrating ecological restoration and conservation biology: A case study for ponderosa pine ecosystems of the Southwest. *Restoration Ecology* 14:4–10.

Noss, R. F., and A. Y. Cooperrider. 1994. *Saving Nature's Legacy: Protecting and Restoring Biodiversity.* Island Press, Washington, D.C.

Noss, R. F., and D. B. Lindenmayer. 2006. The ecological effects of salvage harvesting after natural disturbance—Introduction. *Conservation Biology* 20:946–48.

Odion, D. C., E. Frost, J. R. Strittholt, H. Jiang, D. A. DellaSala, and M. A. Moritz. 2004. Patterns of fire severity and forest conditions in the western Klamath Mountains California. *Conservation Biology* 18:927–36.

Ogden, J., G. H. Stewart, and R. B. Allen. 1996. Ecology of New Zealand *Nothofagus* forests. In *The Ecology and Biogeography of Nothofagus Forests*, T. T. Veblen, R. S. Hill, and J. Read, 25–82. Yale University Press, New Haven, Connecticut.

Oleskog, G., H. Grip, U. Bergsten, and K. Sahlén. 2000. Seedling emergence of *Pinus sylvestris* in characterized seedbed substrates under different moisture conditions. *Canadian Journal of Forest Research* 30:1766–77.

Oliver, C. D., and B. C. Larson. 1996. *Forest Stand Dynamics, Update Edition.* McGraw-Hill, New York.

Orwig, D. A., and D. R. Foster, 1998. Forest response to the introduced hemlock woolly adelgid in southern New England, USA. *Journal of the Torrey Botanical Society* 125:60–73.

Orwig, D. A., and D. B. Kittredge. 2005. Silvicultural options for managing hemlock forests threatened by hemlock woolly adelgid. In *Proceedings of the 3rd Symposium on Hemlock Woolly Adelgid in the Eastern United States*, B. Onken and R. Reardon, 212–17. USDA Forest Service, Forest Health Technology Enterprise Team, Morgantown, West Virginia.

Orwig, D. A., D. R. Foster, and D. L. Mausel. 2002. Landscape patterns of hemlock decline in New England due to the introduced hemlock woolly adelgid. *Journal of Biogeography* 29:1475–87.

O'Shaughnessy, P., and J. Jayasuriya. 1991. Managing the ash-type forest for water production in Victoria. In *Forest Management in Australia*, F. H. McKinnell, E. R. Hopkins, and J. E. D. Fox, 341–63. Surrey Beatty and Sons, Chipping Norton, Australia.

Ough, K. 2002. Regeneration of wet forest flora a decade after clear-felling or wildfire—Is there a difference? *Australian Journal of Botany* 49:645–64.

Ough, K., and A. Murphy. 1998. *Understorey Islands: A Method of Protecting Understorey Flora during Clearfelling Operations*. Internal VSP Report No. 29. Department of Natural Resources and Environment, Melbourne, Australia.

Pahl-Wostl, C. 1995. *The Dynamic Nature of Ecosystems: Chaos and Order Entwined*. Wiley, New York.

Paine, R. T., M. J. Tegner, and E. A. Johnson. 1998. Compounded perturbations yield ecological surprises. *Ecosystems* 1:535–45.

Parminter, J. 1998. Natural disturbance ecology. In *Conservation Biology Principles for Forested Landscapes*, J. Voller and S. Harrison, 3–41. UBC Press, Vancouver, Canada.

Parr, C. L., and A. N. Andersen. 2006. Patch mosaic burning for biodiversity conservation: A critique of the pyrodiversity paradigm. *Conservation Biology* 20:1610–19.

Passovoy, A. D., and Fule, P. Z. 2006. Snag and woody debris dynamics following severe wildfires in northern Arizona ponderosa pine forests. *Forest Ecology and Management* 223:237–46.

Patriquin, M. N., S. Heckbert, C. Nickerson, M. M. Spence, and W. A. White. 2005. *Regional Economic Implications of the Mountain Pine Beetle Infestation in the Northern Interior of British Columbia*. Mountain Pine Beetle Initiative Working Paper 2005-03. Pacific Forestry Centre, Canadian Forest Service, Victoria, Canada.

Peart, D. R., C. V. Coghill, and P. A. Palmiotto. 1992. Effects of logging history and hurricane damage on canopy structure in a northern hardwoods forest. *Bulletin of the Torrey Botanical Club* 119:29–38.

Pedersen, L. 2004. How serious is the mountain pine beetle problem from a timber supply perspective? In *Mountain Pine Beetle Symposium: Challenges and Solutions, October 30–31, 2003, Kelowna, British Columbia, Canada*, Information Report BC-X-399, T. L. Shore, J. E. Brooks, and J. E. Stone, 10–18. Pacific Forestry Centre, Canadian Forest Service, Victoria, Canada.

Perera, A. H., L. J. Buse, and M. G. Weber, editors. 2004. *Emulating Natural Forest Landscape Disturbances. Concepts and Applications*. Columbia University Press, New York.

Perry, D. A. 1994. *Forest Ecosystems*. Johns Hopkins Press, Baltimore, Maryland.

Perry, D. A., M. P. Amaranthus, J. G. Borchers, S. L. Borchers, and R. E. Brainerd. 1989. Bootstrapping in ecosystems. *BioScience* 39:230–37.

Peterson, C. J., and S. T. Pickett. 1995. Forest reorganization: A case study in an old-growth forest catastrophic blowdown. *Ecology* 76:763–74.

Peterson, G. D., G. S. Cumming, and S. R. Carpenter. 2003. Scenario planning: A tool for conservation in an uncertain world. *Conservation Biology* 17:358–66.

Petty, S. J. 1996. *Reducing Disturbance to Goshawks during the Breeding Season.* Forestry Commission Research Note No. 267. Forestry Commission, Edinburgh.

Petty, S. J., G. Shaw, and D. I. Anderson. 1994. Value of nest boxes for population studies and conservation of owls in coniferous forests in Britain. *Journal of Raptor Research* 28:134–42.

Pharo, E., D. B. Lindenmayer, and N. Taws. 2004. The response of bryophytes to landscape context: A large-scale fragmentation study. *Journal of Applied Ecology* 41:910–21.

Phillips, I. D., T. P. Cobb, J. R. Spence, and R. M. Brigham. 2006. Salvage logging, edge effects, and carabid beetles: Connections to conservation and sustainable forest management. *Environmental Entomology* 35:950–57.

Pickett, S. T. A. 1989. Space-for-time substitution as an alternative to long-term studies. In *Longterm Studies in Ecology: Approaches and Alternatives*, G. E. Likens, 110–35. Springer-Verlag, New York.

Pickett, S. T. A., V. T. Parker, and P. Fielder. 1992. The new paradigm in ecology: Implications for conservation biology above the species level. In *Conservation Biology: The Theory and Practice of Nature Conservation*, P. Fielder and S. Jain, 65–88. Chapman and Hall, New York.

Pickett, S. T. A., and J. H. Thompson. 1978. Patch dynamics and the design of nature reserves. *Biological Conservation* 13:27–37.

Pittock, A. B. 2005. *Climate Change: Turning Up the Heat.* CSIRO Publishing, Melbourne, Australia.

Platt, W. J., and J. H. Connell. 2003. Natural disturbances and directional replacement of species. *Ecological Monographs* 73:507–22.

Pousette, J., and C. Hawkins. 2006. An assessment of critical assumptions supporting the timber supply modeling for mountain-pine-beetle-induced allowable annual cut uplift in the Prince George Timber Supply Area. *British Columbia Journal of Ecosystems and Management* 7:93–104.

Preisser, E. L., A. G. Lodge, D. A. Orwig, and J. S. Elkinton. 2007. Range expansion and population dynamics of co-occurring invasive herbivores. *Biological Invasions* 10:201–13.

Prentice, I. C., W. Cramwe, S. P. Harrison, R. Leemans, R. A. Monserud, and A. M. Solomon. 1992. A global biome model based on plant physiology and dominance, soil properties and climate. *Journal of Biogeography* 19:117–34.

Prestemon, J. P., and T. P. Holmes. 2000. Timber price dynamics following a natural catastrophe. *American Journal of Agricultural Economics* 82:145–60.

Prestemon, J. P., and T. P. Holmes, 2004. Market dynamics and optimal timber salvage after a natural disturbance. *Forest Science* 50:495–511.

Prestemon, J. P., D. N. Wear, F. J. Stewart, and T. P. Holmes. 2006. Wildfire, timber salvage, and the economics of expediency. *Forest Policy and Economics* 8:312–22.

Purdon, M., S. Brais, and Y. Bergeron. 2004. Initial response of understorey vegetation to fire severity and salvage logging in the southern boreal forest of Quebec. *Applied Vegetation Science* 7:49–60.

Purdon, M., J. Noël, A. Nappi, P. Drapeau, B. Harvey, S. Brais, Y Bergeron, S. Gauthier, and D. Greene. 2002. *The impact of salvage-logging after wildfire in the boreal forest: les-*

sons from the Abitibi. 4th Research Note. NSERC-UQAT-UQAM Industrial Chair in Sustainable Forest Management, Université du Québec en Abitibi-Témiscamingue, Rouyn-Noranda, Canada. Available on-line at http://inr.oregonstate.edu/download/ impact_of_salvage-logging_after_wildfire_in_boreal_forests.pdf [accessed 31 March 2007].

Putz, F. E., K. H. Redford, J. G. Robinson, R. Fimbel, and G. M. Bate. 2000. *Biodiversity Conservation in the Context of Tropical Forest Management*. World Bank Environment Department Papers. Paper No. 75. Biodiversity Series—Impact Studies. The World Bank, Washington, D.C.

Quine, C. P., J. W. Humphrey, K. Purdy, and D. Ray. 2002. An approach to predicting the potential forest composition and disturbance regime for a highly modified landscape: A pilot study of Strathdon in the Scottish Highlands. *Silva Fennica* 36:233–47.

Rab, M. A. 1998. Rehabilitation of snig tracks and landings following logging of *Eucalyptus regnans* forests in the Victorian Central Highlands—A review. *Australian Forestry* 61:103–13.

Rackham, O. 2001. *Trees and Woodland in the British Landscape*. Phoenix Press, London.

Radeloff, V. C., D. J. Mladenoff, and M. S. Boyce. 2000. Effects of interacting disturbances on landscape patterns: Budworm defoliation and salvage logging. *Ecological Applications* 10:233–47.

Raffa, K. F. 1988. The mountain pine beetle in western North America. In *Dynamics of Forest Insect Populations: Patterns, Causes, Implications*, A. A. Berryman, 505–30. Plenum Press, New York.

Redding, T., R. Winkler, R. Pike, D. Davis, and K. Schorder, compilers. 2007. Mountain pine beetle and watershed hydrology workshop: Preliminary results of research from BC, Alberta and Colorado. FORREX Forestry Research Extension Partnership, Kamloops, Canada. Available on-line at http://www.forrex.org/program/water/ PDFs/Workshops/mpb/MPB-Hydrology_Workshop_Handbook.pdf [accessed 30 September 2007].

Redford, K H. 1992. The empty forest. *BioScience* 42:412–22.

Reeves, G. H., P. A. Bisson, B. E. Rieman, and L. E. Benda. 2006. Post-fire logging in riparian areas. *Conservation Biology* 21:994–1004.

Reice, S. R. 2001. *The Silver Lining: The Benefits of Natural Disasters*. Princeton University Press, Princeton, New Jersey.

Resource Assessment Commission. 1991. *Forest and Timber Inquiry, Draft Report. Volume 1*. Australian Government Publishing Service, Canberra, Australia.

Rice, M. D., B. G. Lockaby, J. A. Stanturf, and B. D. Keeland. 1997. Woody debris decomposition in the Atchafalaya River Basin of Louisiana following hurricane disturbance. *Soil Science Society of America Journal* 61:1264–74.

Roberts, M. R. 2004. Response of the herbaceous layer to natural disturbance in North American forests. *Canadian Journal of Botany* 82:1273–83.

Roberts, M. R. 2007. A conceptual model to characterize disturbance severity in forest harvests. *Forest Ecology and Management* 242:58–64.

Robichaud, P., L. MacDonald, J. Freeouf, D. Neary, D. Martin, and L. Ashmun. 2003. *Postfire Rehabilitation of the Hayman Fire*. General Technical Report RMRS-GTR-114. USDA Forest Service, Rocky Mountain Research Station, Ogden, Utah.

Robinson, G., and J. Zappieri. 1999. Conservation policy in time and space: Lessons from divergent approaches to salvage logging on public lands. *Conservation Ecology* 3(1):3. Available on-line at http://www.ecologyandsociety.org/vol3/iss1/art3/

Rogers, C. S. 1993. Hurricanes and coral reefs: The intermediate disturbance hypothesis revisited. *Coral Reefs* 12:127–37.

Romme, W. H., L. Bohland, C. Perichetty, and T. Caruso. 1995. Germination ecology of some common forest herbs in Yellowstone National Park, Wyoming, USA. *Arctic and Alpine Research* 27:407–12.

Romme, W. H., J. Clement, J. Hicke, D. Kulakowski, L. H. MacDonald, T. L. Schoennagel, and T. T. Veblen. 2006. *Recent Forest Insect Outbreaks and Fire Risk in Colorado Forests: A Brief Synthesis of Relevant Research*. Colorado Forest Restoration Institute, Colorado State University, Fort Collins, Colorado.

Romme, W. H., M. G. Turner, and R. H. Gardner. 1997. A rare episode of sexual reproduction in aspen (*Populus tremuloides*) following the 1988 Yellowstone fires. *Natural Areas Journal* 17:17–25.

Rose, C., B. G. Marcot, T. K. Mellen, J. L. Ohmann, K. Waddell, D. Lindley, and B. Schreiber. 2001. Decaying wood in Pacific Northwest forests: Concepts and tools for habitat management. In *Wildlife–Habitat Relationships in Oregon and Washington*, D. Johnson and T. O'Neil, 580–623. Oregon State University Press, Corvallis.

Rowan, C. A., S. J. Mitchell, and T. Hailemariam. 2002. Effectiveness of clearcut edge windfirming treatments in coastal British Columbia: Short term results. *Forestry* 76:55–65.

Rowe, J. S., and G. W. Scotter. 1973. Fire in the boreal forest. *Quaternary Research* 3:444–64.

Rübsamen, K., I. D. Hume, W. J. Foley, and U. Rübsamen. 1984. Implications of the large surface area to body mass ratio on the heat balance of the greater glider (*Petauroides volans*: Marsupialia). *Journal of Comparative Physiology* 154:105–11.

Rülcker, C., P. Angelstam, and P. Rosenberg. 1994. Natural forest-fire dynamics can guide conservation and silviculture in boreal forests. *SkogForsk* 2:1–4.

Runkle, J. R. 1981. Gap regeneration in some old-growth forests of the eastern United States. *Ecology* 62:1041–51.

Russell, R. E., V. A. Saab, J. G. Dudley, and J. J. Rotella. 2006. Snag longevity in relation to wildfire and postfire salvage logging. *Forest Ecology and Management* 232:179–87.

Russell-Smith, J., P. J. Whitehead, G. D. Cook, and J. L. Hoare. 2003. Response of *Eucalyptus*-dominated savanna to frequent fires: Lessons from Munmarlary, 1973–1996. *Ecological Monographs* 73:349–75.

Saab, V. J., and J. Dudley. 1998. *Responses of Cavity-nesting Birds to Stand-replacement Fire and Salvage Logging in Ponderosa Pine/Douglas-Fir Forests of Southwestern Idaho*. Research Paper RMRS-RP-11. USDA Forest Service, Rocky Mountain Research Station, Ogden, Utah.

Saab, V., R. E. Russell, and J. Dudley. 2007. Nest densities of cavity-nesting birds in relation to post-fire salvage logging and time since fire. *The Condor* 109:97–108.

Sader, S., M. Hoppus, J. Metzler, and S. M. Jin. 2005. Perspectives of Maine forest cover change from Landsat imagery and Forest Inventory Analysis (FIA). *Journal of Forestry* 103:299–303.

Safranyik, L., and A. L. Carroll. 2006. The biology and epidemiology of the mountain pine beetle in lodgepole pine forests. In *The Mountain Pine Beetle: A Synthesis of Biology, Management, and Impacts on Lodgepole Pine*, L. Safranyik and W. R. Wilson, 3–66. Pacific Forestry Centre, Canadian Forest Service, Victoria, Canada.

Safranyik, L., D. M. Shrimpton, and H. S. Whitney. 1974. *Management of Lodgepole Pine to Reduce Losses from the Mountain Pine Beetle*. Forestry Technical Report 1, updated 2002. Pacific Forestry Centre, Canadian Forest Service, Victoria, Canada.

Safranyik, L., and B. Wilson, editors. 2006. *The Mountain Pine Beetle: A Synthesis of Biology, Management, and Impacts on Lodgepole Pine*. Pacific Forestry Centre, Canadian Forest Service, Natural Resources Canada, Victoria, Canada.

Saint-Germain, M., P. Drapeau, and C. Hébert. 2004. Comparison of Coleoptera assemblages from recently burned and unburned black spruce forests of northeastern North America. *Biological Conservation* 118:583–92.

Samuelsson, J., L. Gustafsson, and T. Ingelög. 1994. *Dying and Dead Trees—A Review of Their Importance for Biodiversity*. Swedish Threatened Species Unit, Uppsala, Sweden.

Santelmann, M. V., D. White, K. Freemark, J. I. Nassauer, J. M. Eilers, K. B. Vache, B. J. Danielson, R. C. Corry, M. E. Clark, S. Polasky, R. M. Cruse, J. Sifneos, H. Rustigian, C. Coiner, J. Wu, and D. Debinski. 2004. Assessing alternative futures for agriculture in Iowa, USA. *Landscape Ecology* 19:357–74.

Saveland, J. M., and D. D. Wade. 1991. *Fire Management Ramifications of Hurricane Hugo*. Presented at the 11th Conference on Fire and Forest Meteorology, April 11–19, 1991, Missoula, Montana. Available on-line at http://www.fs.fed.us/rm/pubs/saveland/fire_hugo.pdf [accessed 8 Feb. 2007].

Savill, P. S. 1983. Silviculture in windy climates. *Commonwealth Forestry Bureau* 44:473–88.

Schelhaas, M.-J., G.-J. Nabuurs, and A. Schuck. 2003. Natural disturbances in the European forests in the 19th and 20th centuries. *Global Change Biology* 9:1620–33.

Schelhaas, M.-J., S. Varis, and A. Schuck. 2001. Database on forests disturbances in Europe (DFDE). European Forest Institute, Joensuu, Finland. Available on-line at http://www.efi.fi/projects.dfde/

Schmidt, J. M., and R. H. Frye. 1977. *Spruce Beetle in the Rockies*. General Technical Report RM-49. Rocky Mountain Forest and Range Experiment Station, Fort Collins, Colorado.

Schmiegelow, F. K. A., D. P. Stepnisky, C. A. Stambaugh, and M. Koivula. 2006. Reconciling salvage logging of boreal forests with a natural disturbance management model. *Conservation Biology* 20:971–83.

Schnitzler, A., and F. Borlea. 1998. Lessons from natural forests as key for sustainable management and improvement of naturalness in managed broadleaved forests. *Forest Ecology and Management* 109:293–303.

Schoener, T. W., D. A. Spiller, and J. B. Losos. 2004. Variable ecological effects of hurricanes: The importance of seasonal timing for survival of lizards on Bahamian islands. *Proceedings of the National Academy of Sciences* 101:177–81.

Schuler, T. M., P. Brose, and R. L. White. 2005. *Residual Overstory Density Affects Survival and Growth of Sheltered Oak Seedlings on the Allegheny Plateau*. Research Paper NE-728. USDA Forest Service, Northeastern Research Station, Newtown Square, Pennsylvania.

Schullery, P. 1989. The fires and fire policy. *BioScience* 39:686–94.

Schulte, L. A., R. J. Mitchell, M. L. Hunter, J. F. Franklin, R. K. McIntyre, and B. J. Pallik. 2006. Evaluating the conceptual tools for forest biodiversity conservation and their implementation in the U. S. *Forest Ecology and Management* 232:1–11.

Schulte, L. A., and D. J. Mladenoff. 2005. Severe wind and fire regimes in northern forests: Historical variability at the regional scale. *Ecology* 86:431–45.

Schurbon, J. M., and J. E. Fauth. 2003. Effects of prescribed burning on amphibian diversity in a southeastern US National Forest. *Conservation Biology* 17:1338–49.

Schwilk, D. W., J. E. Keeley, and W. J. Bond. 1997. The intermediate disturbance hypothesis does not explain fire and diversity pattern in fynbos. *Plant Ecology* 132:77–84.

Scott, G. A. 1985. *Southern Australian Liverworts.* Bureau of Flora and Fauna, Australian Government Publishing Service, Canberra, Australia.

Sessions, J., P. Bettinger, R. Buckman, M. Newton, and J. Hamann. 2004. Hastening the return of complex forests following fire: The consequences of delay. *Journal of Forestry* 102(3):38–45.

Sewell, E. K., and J. H. Brown. 1995. Regeneration patterns in low-site oak and oak-pine stands after gypsy moth defoliation and salvage cutting. *Northern Journal of Applied Forestry* 12:109–14.

Seymour, R. S., and M. L. Hunter. 1999. Principles of ecological forestry. In *Maintaining Biodiversity in Forest Ecosystems.* M. L. Hunter, 22–61. Cambridge University Press, Cambridge.

Shakesby, R. A., D. J. Boakes, and C. Coelho. 1996. Limiting the soil degradational impacts of wildfire in pine and eucalyptus forests in Portugal: A comparison of alternative post-fire management practices. *Applied Geography* 16:337–55.

Shakesby, R. A., C. Coelho, and A. D. Ferreira. 1993. Wildfire impacts on soil erosion and hydrology in wet mediterranean forest, Portugal. *International Journal of Wildland Fire* 3:95–110.

Sharma, D. C. 2005. Tsunami damages Indian island's ecology. *Frontiers in Ecology and Environment* 3:5.

Shatford, J. P., D. E. Hibbs, and K. J. Puettmann. 2007. Conifer regeneration after forest fire in the Klamath-Siskiyous: How much, how soon? *Journal of Forestry* 105: 139–46.

Shiel, D., and F. R. Burslem. 2003. Disturbing hypotheses in tropical forests. *Trends in Ecology and Evolution* 18:18–26.

Shinneman, D. J., and W. L. Baker. 1997. Nonequilibrium dynamics between catastrophic disturbances and old growth forests in ponderosa pine landscapes of the Black Hills. *Conservation Biology* 11:1276–88.

Shore, T. L., J. E. Brooks, and J. E. Stone, editors. 2003. *Mountain Pine Beetle Symposium: Challenges and Solutions.* Information Report BC-X-399. Pacific Forestry Centre, Canadian Forest Service and Natural Resources Canada, Victoria, Canada.

Simberloff, D. 1998. Flagships, umbrellas, and keystones: Is single-species management passe in the landscape era. *Biological Conservation* 83:247–57.

Simberloff, D. 2004. Community ecology: Is it time to move on? *American Naturalist* 163:787–99.

Simon, J., S. Christy, and J. Vessels. 1994. Clover-Mist Fire recovery: A forest management response. *Journal of Forestry* 92(11):41–44.

Sinton, D. S., J. A. Jones, J. L. Ohmann, and F. J. Swanson. 2000. Windthrow disturbance, forest composition, and structure in the Bull Run Basin, Oregon. *Ecology* 81:2539–56.

Smith, R. B., and P. Woodgate 1985. Appraisal of fire damage for timber salvage by remote sensing in mountain ash forests. *Australian Forestry* 48:252–63.

Smucker, K. M., R. L. Hutto, and B. M. Steele. 2005. Changes in bird abundance after wildfire: Importance of fire severity and time since fire. *Ecological Applications* 15:1535–49.

Society for Conservation Biology Scientific Panel on Fire in Western U. S. Forests. 2005. *Ecological Science Relevant to Management Policies for Fire-prone Forests of the Western United States.* Society for Conservation Biology, Arlington, Virginia.

Soderquist, T. R., and R. MacNally. 2000. The conservation value of mesic gullies in dry forest landscapes: mammal populations in the box-ironbark ecosystem of southern Australia. *Biological Conservation* 93:281–91.

Sondell, J. 2006. *Knowledge Gained from Operation Gudrun. Resultat No. 7.* Skogforsk, Uppsala, Sweden. (In Swedish with English summary). Available on-line at http://www.skogforsk.se/templates/sf_Product____19169.aspx?sm=1&cpi=4008&ci=46

Sousa, W. P. 1984. The role of disturbance in natural communities. *Annual Review of Ecology and Systematics* 15:353–91.

Spies, T. A., M. A. Hemstrom, A. Youngblood, and S. Hummel. 2004. Conserving old-growth forest diversity in disturbance-prone landscapes. *Conservation Biology* 20:351–62.

Spies, T. A., B. C. McComb, R. S. Kennedy, M. T. McGrath, K. Olsen, and R. J. Pabst. 2007. Potential effects of forest policies on terrestrial biodiversity in a multi-ownership province. *Ecological Applications* 17:48–65.

Spies, T. A., and M. G. Turner. 1999. Dynamic forest mosaics. In *Managing Biodiversity in Forest Ecosystems*, M. Hunter III, 95–160. Cambridge University Press, Cambridge.

Spittlehouse, D. L., and R. B. Stewart. 2003. Adaptation to climate change in forest management. *British Columbia Journal of Ecosystems and Management* 4:1–11.

Stadler, B., T. Müller, and D. Orwig. 2006. The ecology of energy and nutrient fluxes in hemlock forests invaded by hemlock woolly adelgid. *Ecology* 87:1792–1804.

Stadt, J. 2001. *The Ecological Role of Beetle-Killed Trees: A Review of Salvage Impacts.* British Columbia Ministry of Water, Land and Air Protection, Skeena Region, Burns Lake, Canada. Available on-line at http://www.for.gov.bc.ca/hfd/library/documents/bib47094.pdf [accessed 7 February 2007].

Stankey, G. H., B. T. Bormann, C. Ryan, B. Shindler, V. Sturtevant, R. N. Clark, and C. Philpot. 2003. Adaptive management and the Northwest Forest Plan: Rhetoric and reality. *Journal of Forestry* 101(1):40–46.

Stanturf, J. A., S. L. Goodrick, and K. W. Outcalt. 2007. Disturbance and coastal forests: A strategic approach to forest management in hurricane impact zones. *Forest Ecology and Management* 250:119–35.

Stearns, F., and G. E. Likens. 2002. One hundred years of recovery of a pine forest in northern Wisconsin. *American Midland Naturalist* 148:2–19.

Stocks, B. J., M. E. Alexander, B. M. Wotton, C. N. Stefner, M. D. Flannigan, S. W. Taylor, N. Lavoie, J. A. Mason, G. R. Hartley, M. E. Maffey, G. N. Dalrymple, T. W. Blake, M. G. Cruz, and R. A. Lanoville. 2004. Crown fire behaviour in a northern jack pine–black spruce forest. *Canadian Journal of Forest Research* 348:1548–60.

Stocks, B. J., J. A. Mason, J. B. Todd, E. M. Bosch, B. M. Wotton, B. D. Amiro, M. D. Flannigan, K. G. Hirsch, K. A. Logan, D. L. Martell, and W. R. Skinner. 2002. Large forest fires in Canada, 1959–1997. *Journal of Geophysical Research* 108:D1: FFR5, 1–12.

Stokstad, E. 2006. Academic conduct—University bids to salvage reputation after flap over logging paper. *Science* 312:1288.

Stone, R. 1993. Spotted owl plan kindles debate on salvage logging. *Science* 261:287.

Stuart, J. D., M. C. Grifantini, and L. Fox. 1993. Early successional pathways following wildfire and subsequent silvicultural treatment in Douglas-fir hardwood forests, NW California. *Forest Science* 39:561–72.

Stuart-Smith, A. K., J. P Hayes, and J. Schieck. 2006. The influence of wildfire, logging and residual tree density on bird communities in the northern Rocky Mountains. *Forest Ecology and Management* 231:1–17.

Swanson, J. F., S. L. Johnson, S. V. Gregory, and S. A. Acker. 1998. Flood disturbance in a forested mountain landscape. *BioScience* 48:681–89.

Swanson, F. J., J. A. Jones, D. O. Wallin, and J. H. Cissel. 1994. Natural variability—Implications for ecosystem management. In *Eastside Forest Ecosystem Health Assessment, Vol. II, Ecosystem Management: Principles and Applications*, M. E. Jensen, and P. S. Bourgeron, 80–94. General Technical Report PNW-GTR-318. USDA Forest Service, Northwest Research Station, Portland, Oregon.

Syphard, A. D., J. Franklin, and J. E. Keeley. 2006. Simulating the effects of frequent fire on southern California coastal shrublands. *Ecological Applications* 16:1744–56.

Syrjänen, K., R. Kalliola, A. Puolasmaaa, and J. Mattsson. 1994. Landscape structure and forest dynamics in subcontinental Russian European Taiga. *Annales Zoologici Fennici* 31:19–34.

Taylor, R. J., S. L. Bryant, D. Pemberton, and T. W. Norton. 1985. Mammals of the Upper Henty River region, western Tasmania. *Papers and Proceedings of the Royal Society of Tasmania* 119:7–15.

Taylor, S. W., A. L. Carroll, R. I. Alfaro, and L. Safranyik. 2006. Forest, climate and mountain pine beetle outbreak dynamics in western Canada. In *The Mountain Pine Beetle: A Synthesis of Biology, Management, and Impacts on Lodgepole Pine*, L. Safranyik, and W. R. Wilson, 67–94. Pacific Forestry Centre, Canadian Forest Service, Victoria, Canada.

Taylor, S. W., J. Parminter, and G. Thandi. 2005. *Logistic Regression Models of Wildfire Probability in British Columbia.* Annual Technical Report Supplement 2, Forest Science Program Project Y05-01233. Pacific Forestry Centre, Canadian Forest Service, Victoria, Canada.

Tebo, M. E. 1985. *The Southeastern Piney Woods: Describers, Destroyers, Survivors.* M. A. Thesis, Florida State University, Tallahassee, Florida.

Thompson, J. R., T. A. Spies, and L. M. Ganio. 2007. Re-burn severity in managed and unmanaged vegetation in the Biscuit Fire. *Proceedings of the National Academy of Sciences* 104:10743–48.

Thrower, J. 2005. Earth Island Institute versus United States Forest Service: Salvage logging plans in Star Fire region undermine Sierra Nevada framework. *Ecology Law Quarterly* 32:721–28.

Titus, J. H., and E. Householder. 2007. Salvage logging and replanting reduce understorey cover and richness compared to unsalvaged-unplanted sites at Mount St. Helens, Washington. *Western North American Naturalist* 67:219–31.

Trabaud, L. 2003. Towards a sounder fire ecology. *Frontiers in Ecology and the Environment* 1:274–75.

Tremblay, M., C. Messier, and D. J. Marceau. 2005. Analysis of deciduous tree species dynamics after a severe ice storm using SORTIE model simulations. *Ecological Modeling* 187:297–313.

Trombulak, S. C., and C. A. Frissell. 2000. Review of ecological effects of roads on terrestrial and aquatic communities. *Conservation Biology* 14:18–30.

Turner, M. G., W. L. Baker, C. J. Peterson, and R. K. Peet. 1998. Factors influencing succession: Lessons form large, infrequent natural disturbances. *Ecosystems* 1:511–23.

Turner, M. G., W. H. Romme, R. H. Gardner, and W. W. Hargrove. 1997. Effects of fire size and pattern on early succession in Yellowstone National Park. *Ecological Monographs* 67:411–33.

Turner, M. G., W. H. Romme, and D. B. Tinker. 2003. Surprises and lessons from the 1988 Yellowstone fires. *Frontiers in Ecology and Environment* 1:351–58.

Ulanova, N. G. 2000. The effects of windthrow on forests at different spatial scales: A review. *Forest Ecology and Management* 135:155–67.

Ulbricht, R., A. Hinrichs, and Y. Ruslim. 1999. *Technical Guideline for Salvage Felling in Rehabilitation Areas after Forest Fires. Report 1 of the Sustainable Forest Management Project.* Samarinda, Indonesia.

United States Government Accountability Office. 2006. *Biscuit Fire Recovery Project. Analysis of Project Development, Salvage Sales, and Other Activities.* U.S. Government Accountability Office, Washington, D.C.

USDA Forest Service. 1996. *Hurricane Hugo: South Carolina Forest Land Research and Management Related to the Storm.* General Technical Report SRS-005. USDA Forest Service, Ashville, North Carolina.

USDA Forest Service. 2001. *Environmental Assessment: Opal Creek Scenic Recreation District, Willamette National Forest.* USDA Forest Service, Pacific Northwest Region, Mill City, Oregon. www.us.fed.us/r6/pgr/afterfire/assessment/index.html

USDA Forest Service. 2004. *Record of Decision for the Rodeo-Chediski Fire Salvage Project.* USDA Forest Service, Apache-Sitgreaves National Forest, Sringerville, Arizona.

USDA Forest Service. 2006. *Bear Tornado Recovery Project: Environmental Assessment.* Council Ranger District, USDA Forest Service, Council, Idaho. Available on-line at http://www.fs.fed.us/r4/payette/projects/bear_blow_down/bear_ea.shtml [accessed 28 March 2008].

Van der Rhee, R., and R. H. Loyn. 2002. The influence of time since fire and distance from fire boundary on the distribution of arboreal marsupials in the *Eucalyptus regnans*-dominated forest in the Central Highlands of Victoria. *Wildlife Research* 29:151–58.

Van Lear, D. H., J. E. Douglass, S. K. Cox, and M. K. Augspurger. 1985. Sediment and nutrient export in runoff from burned and harvested pine watersheds in the South Carolina piedmont. *Journal of Environmental Quality* 14:169.

Van Nieuwstadt, M. G., D. Shiel, and K. Kartawinata. 2001. The ecological consequences of logging in the burned forests of east Kalimantan, Indonesia. *Conservation Biology* 15:1183–86.

Veblen, T. T. 2000. Disturbance patterns in southern Rocky Mountain forests. In *Forest Fragmentation in the Southern Rocky Mountains*, R. L. Knight, F. W. Smith, S. W. Buskirk, W. H. Romme, and W. L. Baker, 31–54. Island Press, Washington, D.C.

Veblen, T. T., C. Donoso, T. Kitzberger, and A. J. Rebertus. 1996. Ecology of southern Chilean and Argentinean *Nothofagus* forests. In *The Ecology and Biogeography of Nothofagus Forests*, T. T. Veblen, R. S. Hill, and J. Read, 293–353. Yale University Press, New Haven, Connecticut.

Veblen, T. T., K. S. Hadley, E. M. Nel, T. Kitsberger, M. Reid, and R. Villalba. 1994. Disturbance regime and disturbance interactions in a Rocky Mountain subalpine forest. *Journal of Ecology* 82:125–35.

Vera, F. W. M. 2000. *Grazing Ecology and Forest History*. CABI Publishing, New York.

Victoria Department of Natural Resources and Environment. 1996. *Code of Practice: Code of Forest Practices for Timber Production. Revision No. 2, November 1996*. Department of Natural Resources and Environment, Melbourne, Australia.

Victoria Department of Sustainability and Environment. 2003. *Proposed Salvage Logging Prescriptions for the 2003 Eastern Victoria Fire*. Department of Sustainability and Environment Report, Melbourne, Australia.

Victoria Department of Sustainability and Environment. 2007. *Salvage Harvesting Prescriptions Table*. Department of Sustainability and Environment, Melbourne, Australia.

Wace, N. 1977. Assessment of the dispersal of plant species—The car-borne flora of Canberra. *Proceedings of the Ecological Society of Australia* 10:166–86.

Wade, D. D., J. K. Forbus, and J. M. Saveland, 1993. *Photo Series for Estimating Post-Hurricane Residues and Fire Behavior in Southern Pine*. General Technical Report SE-082. USDA Forest Service, Southeastern Forest Experiment Station, Ashville, North Carolina.

Wadleigh, L., and M. J. Jenkins. 1996. Fire frequency and the vegetative mosaic of a spruce-fir forest in northern Utah. *Great Basin Naturalist* 56:28–37.

Wales, B. C. 2001. The management of insects, diseases, fire and grazing and implications for terrestrial vertebrates using riparian habitats in eastern Oregon and Washington. *Northwest Science* 75(Special Issue S1):119–27.

Walker, A. 1999. *Examination of the Barriers to Movements of Tasmanian Fish*. Honours thesis, School of Zoology, University of Tasmania, Hobart, Australia.

Walker, B. H., C. S. Holling, S. Carpenter, and S. C. Kinzig. 2004. Resilience, adaptability and transformability. *Ecology and Society* 9(2). Available on-line at http://www.ecologyandsociety.org/vol9.iss2/art5

Walker, B., and D. Salt. 2006. *Resilience Thinking: Sustaining Ecosystems and People in a Changing World*. Island Press, Washington, D.C.

Wallace, L. L., editor. 2004. *After the Fires: The Ecology of Change in Yellowstone National Park*. Yale University Press, New Haven, Connecticut.

Wallace, L. L., F. J. Singer, and P. Schullery. 2004. The fires of 1988: A chronology and invitation to research. In *After the Fires: The Ecology of Change in Yellowstone National Park*, L. L. Wallace, 3–9. Yale University Press, New Haven, Connecticut.

Walters, C. J. 1986. *Adaptive Management of Renewable Resources*. Macmillan Publishing Company, New York.

Walters, C. J., and C. S. Holling. 1990. Large scale management experiments and learning by doing. *Ecology* 71:2060–68.

Wardell-Johnson, G., and P. Horowitz. 1996. Conserving biodiversity and the recognition of heterogeneity in ancient landscapes: A case study from south-western Australia. *Forest Ecology and Management* 85:219–38.

Waring, R. H., and G. B. Pitman. 1985. Modifying lodgepole pine stands to change susceptibility to mountain pine beetle attack. *Ecology* 66:889–97.

Watson, P. A. 2006. Impact of the mountain pine beetle on pulp and papermaking. In *The Mountain Pine Beetle: A Synthesis of Biology, Management, and Impacts on Lodgepole Pine*, L. Safranyik and W. R. Wilson, 255–75. Pacific Forestry Centre, Canadian Forest Service, Victoria, Canada.

Weir, J. M. H., E. A. Johnson, and K. Miyanishi. 2000. Fire frequency and the spatial age mosaic of the mixedwood boreal forest in Western Canada. *Ecological Applications* 10:1162–77.

Wells, G. 1998. *The Tillamook: A Created Forest Comes of Age*. Oregon State University Press, Corvalis, Oregon.

Westerling, A. L., H. G. Hidalgo, D. R. Cayan, and T. W. Swetnam. 2006. Warming and earlier spring increase western U. S. forest wildfire activity. *Science* 313:940–43.

Whelan, R. J. 1995. *The Ecology of Fire*. Cambridge University Press, Cambridge.

Whelan, R. J. 2002. Managing fire regimes for conservation and property: An Australian response. *Conservation Biology* 16:1659–61.

Whelan, R., L. Rodgerson, C. R. Dickman, and E. F. Sutherland. 2002. Critical life cycles of plants and animals: Developing a process-based understanding of population changes in fire-prone landscapes. In *Flammable Australia: The Fire Regimes and Biodiversity of a Continent*, R. A. Bradstock, J. E. Williams, and A. M. Gill, 94–124. Cambridge University Press, Melbourne, Australia.

White, I., N. Mueller, T. Daniell, and R. Wasson. 2006. The vulnerability of water supply catchments to bushfires: Impacts of the January 2003 wildfires on the Australian Capital Territory. *Australian Journal of Water Resources* 10:1–15.

White, P. S., and S. T. A. Pickett. 1985. Natural disturbance and patch dynamics: An introduction. In *The Ecology of Natural Disturbance and Patch Dynamics*, S. T. A. Pickett and P. S. White, 3–13. Academic Press, New York.

Whitehead, R. J., and G. L. Russo. 2005. *Beetle-proofed Lodgepole Pine Stands in Interior British Columbia Have Less Damage from Mountain Pine Beetle*. Information Report BC-X-402. Pacific Forestry Centre, Canadian Forest Service, Victoria, Canada.

Wiens, J. A., R. L. Schooley, and R. D. Weekes. 1997. Patchy landscapes and animal movements: Do beetles percolate? *Oikos* 78:257–64.

Wikars, L.-O. 2001. The wood-decaying fungus *Daldinia loculata* (Xylariaceae) as an indicator of fire-dependent insects. *Ecological Bulletins* 49:263–68.

Willis, K. J., and H. J. B. Birks. 2006. What is natural? The need for a long-term perspective in biodiversity conservation. *Science* 314:1261–65.

Wilson, J. B. 1994. The intermediate disturbance hypothesis of species coexistence is based on patch dynamics. *New Zealand Journal of Ecology* 18:176–81.

Wisdom, M. J., M. Vavra, J. M. Boyd, M. A. Hemstrom, A. A. Ager, and B. K. Johnson. 2006. Understanding ungulate herbivory-episodic disturbance effects on vegetation dynamics: Knowledge gaps and management needs. *Wildlife Society Bulletin* 34:283–92.

Woldendorp, G., and R. J. Keenan. 2005. Coarse woody debris in Australian forest ecosystems: A review. *Austral Ecology* 30:834–43.

Woodward, F. L., and B. G. Williams 1987. Climate and plant distribution at global and local scales. *Vegetatio* 69:189–97.

World Commission on Forests and Sustainable Development. 1999. *Our Forests Our Future: Report of the World Commission on Forests and Sustainable Development.* Cambridge University Press, Cambridge.

Yamamoto, S.-I. 1992. The gap theory in forest dynamics. *Botanical Magazine* (Tokyo) 105:375–83.

Yarie, J., L. Viereck, K. van Cleve, and P. Adams. 1998. Flooding and ecosystem dynamics along the Tanana River. *BioScience* 48:690–95.

Zackrisson, O. 1977. Influence of forest fires on the north Swedish boreal forest. *Oikos* 29:22–32.